JOURNEY THROUGH

INVENTIONS

JOURNEY THROUGH

INVENTIONS

RON TAYLOR

HAMLYN

Published in 1991 by
Hamlyn Children's Books,
part of Reed International Books,
Michelin House, 81 Fulham Road,
London SW3 6RB

ISBN 0 600 57117 3

Printed in Italy

CONTENTS

STONE AGE INVENTIONS

An invention is something completely new - something thought-up and designed for a specific job. What were the first inventions? They certainly include stone weapons and tools made by the ancestors of modern humans. Five million or more years ago, these human ancestors, the Australopithecines, roamed the hot plains of Africa. They had faces and brains like chimpanzees and stood only about 1 m high, but unlike apes they walked and ran upright. This made them effective hunters, particularly as it freed their hands for holding stone objects, such as the simple pebble tools they employed to cut up the animals they used for food.

Fire, Spears, and Hand Axes

Much later on - a million or more years ago - early people made a primitive type of spear by charring one end of a stick in fire, then rubbing it to a sharp point. The making and control of fire was itself a great invention of this time, as fire had many important uses such as providing warmth, cooking food, and keeping away wild animals.

Later still, about 200,000 years ago, Stone Age people invented the first hand axes: stone tools or weapons used for chopping and slicing. These were sharper and more precise than the much older pebble tools, and were made by a process known as pressure flaking. A suitable lump of hard stone, such as flint, was chosen, then struck or pressed against another until it split. This process was continued until the hand axe had at least one sharp edge. For chopping and slicing, it was held directly in the hand, without a wooden handle.

Experts in Stone and Bone

Stone hand axes continued to be made over a vast stretch of time - the Old Stone Age - and improved greatly in accuracy and sharpness. By about 30,000 years ago, very finely-shaped stone arrow heads and spear heads were also made by the pressure flaking technique and were then bound on to the ends of wooden shafts.

Elegant bone tools also date from this time onwards. Arrow heads, chisels, and eventually even delicate sewing needles were made from animal bones, horns, and antlers. By about 15,000 years ago, some of these bone tools were decorated with designs or pictures of animals.

Early Artists

People still lived in caves or simple shelters at this time, but otherwise resembled ourselves in many ways. They looked like us, their brains were just as big, and they were just as inventive. They hunted animals with spears, throwing sticks, and bows and arrows. They cooked food and made fur clothing. Most remarkably of all, they were wonderful artists, as cave paintings at Altamira in Spain and Lascaux in France show. They painted pictures of animals they hunted, using colours made from powdered minerals such as black, brown, and red ochre. These were mixed with animal fat to make the colours stick to the walls of the cave.

Left: Some animals use tools. This chimp extracts tasty termites with a stick.

PEBBLE TOOLS INVENTED 5 MILLION YEARS AGO • HAND AXES 200,000 YEARS AGO

Right: Human beings invented picture painting more than 15,000 years ago. This Bushman rock painting of animals dates from more recent times, but is very similar to Old Stone Age cave paintings.

Below: The sharp-edged axe was one of the most important inventions of the Old Stone Age. It was made by a process called pressure flaking.

REFINED STONE AND BONE TOOLS 30,000 YEARS AGO • FIRST ART 15,000 YEARS AGO

FARMING TOOLS

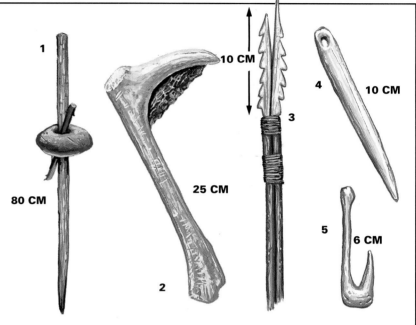

FARMING TOOLS

The first farming tools were made mostly of wood, bone, and stone. The digging stick (**1**) was a sharpened stick with a stone weight to help drive it into the soil. It was the ancestor of both the hoe and the plough. The flint sickle (**2**) used for reaping cereal grasses such as wheat and barley, had a sharp cutting edge of hard flint, wedged into a bone handle, in this case part of a deer's antler. Stone Age fishermen also had a range of finely-made bone tools. These included spear prongs (**3**), needles for making fishing nets (**4**), and fishing hooks (**5**).

80 CM

10 CM

25 CM

10 CM

6 CM

Below: The first houses were simple huts invented by people of the Middle Stone Age, about 12,000 years ago. Their tools included those shown in the foreground and the box above.

PEOPLE BECOME SETTLED FARMERS c.10,000 BC • **PLOUGH INVENTED** c.8000 BC

About 12,000 years ago, people first began to settle down and live in small villages and encampments. They became the first farmers, keeping domestic animals such as goats and pigs and harvesting food cereals: grasses such as barley, oats, millet, and wheat. In time, they cultivated or bred specially improved types of these animals and plants to provide richer food than the wild varieties.

Early Farming Tools

Even before people first settled down, they had, on their nomadic wanderings, gathered or reaped local grasses for food. To do this they invented a sickle, or curved knife, made out of hard flint and held in one hand. Reaping with such stone sickles continued for thousands of years, well into the period when people settled down, the New Stone Age. Of course, farmworkers and gardeners even today still use metal-bladed sickles, for example to trim rough grass patches and hedges.

The first important farming tool actually invented by settled-down man was the digging stick. This was a pointed stick weighed down with a stone to make it sink more easily into hard soil. Some time later the flail was invented - a stiff sort of whip for threshing out food grains from cereal plants. The grains in turn were ground down to make flour for bread, as they still are today.

Ploughs, Hoes, and Scythes

Such primitive tools as flint sickles and digging sticks were the ancestors of later, larger or more elaborate farming inventions. From the flint sickle came the scythe, a larger form of curved knife with a long wooden handle which was held in both hands and swept around to mow grasses. The scythe only really came into its own when metals first began to be beaten out into such shapes as scythe blades and sword blades, that is in the Bronze Age that followed the New Stone Age.

From the digging stick came two major inventions. The hoe is a flattened, lengthened digging stick held in both hands for jabbing into and turning over soil and rooting out weeds. Hoes with metal blades are common tools in gardens today.

An even more important invention is the plough, a digging stick dragged along to turn over long lines of soil. A plough does several useful jobs at the same time. By turning over soil in a field, it aerates it, which is good for plant growth. Also, it buries dead stalks and other leftovers from a previous season's growth, which then rot to provide food for the new season's growth. Finally, it provides long hollows suitable for sowing seeds.

The earliest ploughs were made entirely of wood and dragged along by the farmer himself. Later ploughs were dragged along much more conveniently by domestic animals such as oxen, and later still by horses. Surprisingly, most ploughs were not improved by metal parts for thousands of years. A few iron-bladed ploughs are known to have existed about 1000 BC, but primitive all-wooden ploughs are still used by poor farmers in some countries.

Below: This Bangladeshi farmer uses an ancient type of wooden plough.

THE FIRST BUILDINGS

Human beings were around for a long time before they invented buildings to live in. Our remote ancestors, the Australopithecines of 5-2 million years ago, probably never had a permanent roof over their heads. In their warm African climate, they had no great need of shelter, except when escaping a storm under a tree or in a cave.

By a million or more years ago, people had spread out to live in other parts of the world, many of which had a very cold Ice Age climate. But still they made very few buildings of their own, preferring to live in caves and other natural shelters. When the famous naturalist Charles Darwin visited the southern tip of South America as late as the nineteenth century, he found there the Ona Indians, who went about naked in the cold, blustery climate and made only the most primitive kinds of shelter. These were boards of woven twigs, propped up to afford some protection against the constant cold wind.

Tents and Mud Huts

Many of the world's ancient peoples were nomads or wanderers, driving their cattle from one feeding place to another, sometimes across great distances.

Such nomads still live in the remoter parts of Asia, on the high plains of Tibet, Sinkiang, and Mongolia. They shelter in tents they make from the thick skins of animals such as the yak. It seems likely that nomads of man's remote past also made such tents, but these have all long since rotted away.

When people settled down to live in villages, they built huts with mud, sticks, and grasses, materials they gathered locally. These huts, too, have disappeared almost entirely without trace, although similar ones can still be seen in African villages. In a few cases, however, some evidence remains. In particular, traces of huts and primitive kitchens have been found from the Maglemosian culture, which flourished in Denmark 12,000 years ago.

Right: Towns and cities of Mesopotamia, 5,000 years ago, were built with mud bricks. The large building in the background is a ziggurat.

Left: The step pyramid of King Zoser was the first of the great pyramids of Egypt to be built. It is still standing after 4,800 years.

Bricks, Stores, and Houses

As people prospered as small farmers, they needed to protect their harvested crops against the weather. Mud huts could be used not only for living in but also for storing grain. Larger grain stores were built of mud bricks dried hard in the sun, and were the first brick buildings.

Richer farmers and traders then began to build themselves houses of sun-dried brick. These had several rooms, compared with the mud hut's single one, but hardly any of them have survived, because after some centuries, sun-dried brick crumbles away. Only a few very big brick buildings of those times have left any trace today.

Religious Buildings

One such large, sun-dried brick building is the great ziggurat (stepped pyramid) of Ur in Ancient Mesopotamia, now Iraq. Built over 4,000 years ago, this is still recognizable, though much crumbled. A ziggurat was a religious building, as were many of the giant buildings of ancient times.

Others were built not of brick but of stone. Stonehenge, in Wiltshire, dates from about the same time. It is a double ring of large stones, some of the heaviest of which were transported from long distances away. Stonehenge and other megalithic, or upright-stone, buildings, were used for a form of Sun worship. Egyptian pyramids are the greatest of all the ancient stone buildings. Constructed with amazing precision from huge stone blocks mined elsewhere and floated on rafts down the River Nile, they were the tombs of royalty.

THE FIRST CRAFTS

Crafts were first practised as long ago as the Old Stone Age. In the caves of southern France, the Aurignacian people left behind very fine stone and bone tools, weapons, and carvings dating from as long as 30,000 years ago. They made simple cups and dishes of stone and clay. They also made their own animal-skin clothes. Later cave dwellers, such as the Magdalenian artists of 15,000 years ago, were probably the first weavers of cloth. None of these examples of clothes-making skill has survived.

Looms and Weaving

In the earliest towns, such as Jericho in the Palestine of 6000 BC, a weaving trade was certainly active. Spindles and other parts of weaving looms have been found at the sites of these towns. The first looms were frames pegged out flat on the ground. Later, the more familiar type of upright loom was invented. In both cases, a spindle holding a length of thread is passed between many other, fixed, threads to make woven cloth. Hand-operated looms such as these are still used in many countries today.

Kilns and Pottery

Clay jars and pitchers for holding water date from a very early time in settled history. At first these were made by coiling ropes of clay round and round until they formed a pot of some kind, which was then dried hard in the sun.

Then, in one of the early towns in about 3000 BC, the first potter's wheel was invented. This was simply a round, flat stone, pivoted in the middle and spun by hand. In the centre the potter stuck a lump of clay, which he shaped into an elegant vessel as he spun the wheel. Later, potter's wheels were improved by adding a kick-wheel below the forming-wheel. This let the potter use his feet for turning the wheels and freed both his hands for pot-making. Such spun pots were often too large to be hardened quickly in the sun. Instead, they were fired in a mud-brick furnace or kiln heated up with burning wood.

The Metal Ages

Very occasionally, someone might make use of a lump of metal found lying around - pure or native copper or an iron meteorite. But the skill and craft of metallurgy, forming metals into useful shapes, had to wait until the first civilizations.

The Ancient Egyptians were skilled in making objects from copper, both the native metal and copper they extracted from one of its ores by roasting the ore in a wood or charcoal fire. This period of time, dating from about 3200 BC, is known as the Copper Age. A century or so later, in the cities of Mesopotamia, the Bronze Age began. Bronze is an alloy, or mixture of metals, formed by roasting together ores of the metals copper and tin. Bronze is harder and tougher than copper and found many uses as swords, spears, body armour, and storage vessels. These were beaten into the correct shape from the mass of bronze alloy roasted out or smelted from the metal ores.

Gold is a rare metal always found in its native state. It is also rather soft and easily formed into intricate shapes. The Ancient Egyptians were among the first people to make gold and silver jewellery.

Left and below: The marble "Venus" figure from Ancient Turkey was carved 8-9,000 years ago. The Sumerian gold and lapis lazuli bull is 4,500 years old.

POTTERY AND LOOM-WEAVING INVENTED c.6000 BC • POTTER'S WHEEL c.3000 BC

Above: A busy street in Jericho, 5,000 years ago. Many useful crafts had been invented by this time. Those shown are pottery, weaving and jewellery-making. The chair carriers are on their way to collect an important person.

COPPER AGE FROM 3200 BC • BRONZE AGE 3100 BC • GOLD JEWELLERY 3000 BC

SCALES AND MEASURES

Our early ancestors owned very few belongings that needed to be counted. The first herdsmen and villagers were different: they had animals to be counted and stocks of grain to be measured. Kings and governors of the first cities were even more concerned with what they owned. They left behind records of their property as cuneiform writing scratched on to clay tablets, thousands of which have been found at the sites of ancient cities between the rivers Euphrates and Tigris in what is now Iraq.

Right: A Chinese abacus, with 5 + 2 beads on each wire. The calculations in the box below are done on a Russian abacus, with 10 beads on each wire.

ABACUS ARITHMETIC

ADDITION SUM: 625+94 **1** First, set up 625. **2** Next, add (move down) 4 beads on the units (**U**) wire to make 9 beads. **3** Then try to add (move down) 9 beads on the tens (**T**) wire. But this would make 11 beads. So, add (move down) 1 bead on the hundreds (**H**) wire (which has the same value as 10 beads on the tens wire), and leave 1 on the tens wire, by moving one bead up. This leaves the answer, 719.

SUBTRACTION SUM: 1052-441 **1** First, set up 1052. **2** Next, take away (move up) 1 bead on the units wire and 4 beads on the tens wire. **3** Next, try to take away (move up) 4 beads on the hundreds wire. You cannot do this as there are no beads to move up. Instead, "borrow" 1 bead from the thousands wire by moving it up. This is the same as adding 10 beads to the hundreds wire, so move them all down. You now have 10 beads from which to take away 4, so move 4 beads up. This leaves the answer, 611.

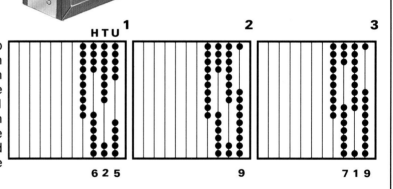

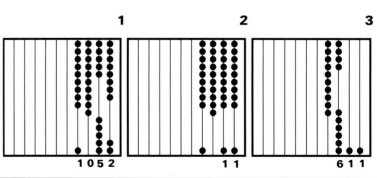

Left: A Mayan calendar from Central America. This shows the passage of time, but not in months or years.

Below: This Roman calendar clearly shows the months of the year.

Arithmetic

To count anything you need a system of arithmetic. Among the first of these to be invented was the decimal system, using multiples of 10. This is hardly surprising, since finger counting was an obvious method for calculating numbers. However, another system of arithmetic invented 5,000 years ago by the Mesopotamians used multiples of 60 instead of 10. We still employ this ancient sexagesimal system to count time in seconds and minutes.

To keep a record of large numbers, people first wrote on clay tablets, and later on, in Ancient Egypt, on papyrus made of reeds from the River Nile. More suitable for the smaller numbers of household property were such methods as knotted cords, notched sticks, and counting stones and sticks. The counting stones were later put on to a frame to make the abacus, still used by shopkeepers in bazaars all across Asia.

Calendars

Another ancient method of counting arose from the ways in which nature repeats itself. Thus, the time from new moon to old moon became a "moonth" or month. Other natural repetitions are the day (sun-up to sundown) and the year, which was first calculated from the way in which positions of the Sun, Moon, and stars in the sky repeat themselves roughly every 365 days. (Roughly, because we need the leap year!) From these calculations the first calendars were invented by the peoples of Ancient Egypt and Mesopotamia.

Scales, Measures, and Money

The first shopkeepers needed to measure out amounts accurately to the first customers, which led to the invention of weighing scales and standard weights. For such purposes as building and land surveying, lengths also needed to be standardized. A very early unit of length was the cubit: the length of the forearm from the tip of the middle finger to the elbow. Of course, since forearms vary, this was not very accurate! Neither, more recently, was the foot until it became standardized.

People first bartered or exchanged goods against goods, but later, money was invented. In our long history, many different objects have been used as money, including seashells, animals and their bones, human skulls, and slaves. Only gradually were metals used for money. "Drachma", a Greek unit of money, probably means "a handful of iron nails". Eventually, in the eighth century BC, metal coins were invented in several countries.

CALENDARS INVENTED c.3000 BC • ABACUS c.2000 BC • METAL COINS c.800 BC

WORDS AND ALPHABETS

Below: Cuneiform writing on a Sumerian clay tablet more than 4,000 years old.

Bottom: A Viking runestone, carved 1,000 years ago, with its strange runic writing.

Long before they developed proper speech, our early human ancestors communicated with one another by grunts and signs. Their ability to use language had developed little beyond that of other higher animals. At this time in our remote past, people almost certainly had not yet invented any form of written language.

After human beings had learned to speak, they did sometimes draw and paint signs and pictures on rocks, to show how something had happened. A cave painting can sometimes be thought of as a picture-message of a general kind, such as "we went hunting antelopes".

Picture Writing

Pictographs, or simplified picture-messages, were still being drawn on tree bark by North American Indians in the nineteenth century. A pictograph might be a simple drawing, say of a bison. Or it might be a bison with three straight marks, meaning perhaps "I have killed three bison". Or it might be a man with all his ribs showing, meaning "Our people are starving". So a pictograph is more like an *idea* than like a single sound or word.

Out of such picture-messages, more modern methods of writing have developed. But between the pictograph and the written word and sentence, there have been many different stages.

Cuneiform Writing

The first proper writing was invented about 3200 BC by a people called the Sumerians, who lived in the world's first cities in Ancient Mesopotamia. This was long before paper was invented, and the Sumerians used instead a very easily obtained local material, clay. To write on hardened clay tablets, they needed to press down much more firmly than we do with pen on paper, so used a sharp marker or stylus to make thin wedge-shaped marks.

"Cuneiform", a Latin word meaning wedge-shaped, is the special name of this writing. In its earliest form it looks very much like pictographic writing. Later on, some of its marks came to represent sounds as well as pictures. For example, the Sumerian word for "hand" sounded something like "ssu", and in time the cuneiform picture of a hand also came to mean this particular sound.

PICTURE-MESSAGES FROM c.20,000 YEARS AGO • PICTOGRAPHS FROM c.8000 BC

CUNEIFORM WRITING

The first forms of writing were simplified pictures of objects, called pictographs. In Ancient Mesopotamia, pictographs were scratched into the surface of clay tablets with a hard, wedge-shaped tool called a stylus. Gradually, through the centuries, the pictographs themselves became simplified into a number of straight stylus marks or characters that no longer looked anything like the original subjects. This is called cuneiform, or wedge-shaped writing. Some of the cuneiform characters stood for objects, such as "bird", "ox", and "orchard". Other cuneiform characters stood for actions, such as the verb "to plough".

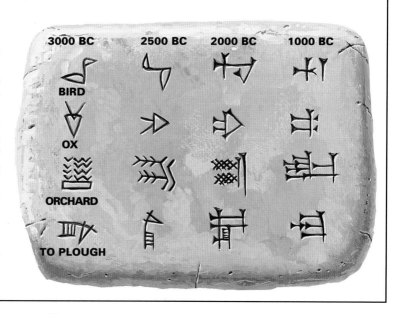

3000 BC	2500 BC	2000 BC	1000 BC
BIRD			
OX			
ORCHARD			
TO PLOUGH			

Left: An open-air school in Egypt, 4,000 years ago. The scroll held by the teacher is papyrus made from reeds of the River Nile.

Chinese writing, which is done with brush on paper, looks very different from cuneiform writing, but also arose in the same manner. Some of its marks, such as that for "man", still look like pictures of what they represent.

Alphabets

The word alphabet comes from "alpha" and "beta", the words for the first two letters of the Greek alphabet. In their turn, the Ancient Greeks got their alphabet from earlier peoples, the Semites. The first two letters of Hebrew, a Semitic language, are "aleph" and "beth".

As in English and most other modern languages, alphabetic letters are no longer pictures of anything, but stand purely for sounds. Single alphabetic letters are divided into vowels, or open sounds, and consonants, or closed sounds. Combinations of these are called syllables. Sometimes, as in "to" and "pi", syllables are also simple words.

CUNEIFORM WRITING INVENTED c.3200 BC • FIRST ALPHABETS INVENTED c.800 BC

IRON, STEEL, AND GLASS

As long ago as 2000 BC, the more advanced peoples of the world, such as those in Ancient Egypt and Mesopotamia, already used a number of metals for weapons, tools, and jewellery. Hardest and strongest of these metals was bronze, an alloy of tin and copper (see also pages 12-13). But even harder and stronger than bronze are iron and steel, the next metals to be used by people.

The Iron Age

About 1400 BC in the Hittite empire, which extended over parts of the countries now called Syria and Turkey, iron tools and weapons first began to be made. However, the wood fires of the Hittite iron-smiths were never hot enough to melt iron or its ores to enable the metal to be poured into moulds, as was done at that time with many bronze articles.

Instead, the Hittite ironsmiths heated iron ores until these were partly converted into a mixture of iron metal and slag or waste, then hammered the hot iron out of the slag and formed it into the required shapes. This method makes a very strong, tough metal called wrought iron, which was ideal for swords and other weapons.

Even stronger is steel, an alloy of iron that contains small amounts of non-metals, most importantly, carbon. Most types of steel have been made

FIRST GLASS OBJECTS MADE c.3000 BC • FIRST IRON OBJECTS MADE c.1400 BC

only in the last few hundred years, using furnaces hot enough to melt iron ores. However some of the Hittite ironsmiths managed to make a type of steel no less than 3,000 years ago! They did this by hammering extra hard and long at the hot iron slag so that some of the carbon from their wood fires combined chemically with the iron metal. This made a form of steel called cementation steel.

Ancient Chinese ironsmiths were also a thousand or more years ahead of their time. About 400 BC, they succeeded in melting iron metal, using charcoal fires fanned to white heat with blasts of air from highly efficient bellows. They poured or cast this molten iron into moulds to make shapes of cast iron. They used this extremely hard, springy form of iron for making articles as different as cooking pots and weapons such as crossbows.

Below: Hittite smiths, 3,000 years ago, beating wrought-iron swords out of iron ore roasted in a wood fire.

Below: Two Egyptian princesses of 3,750 years ago moulded in red glass.

Glass-Making

Glass is another hard material useful for containers and other household objects. It is made by heating together the natural substances sand, lime, and soda to about 1000 °C, when they fuse or melt together to form the transparent substance we know as glass.

The Ancient Egyptians knew where to find these substances and their charcoal fires could just reach the necessary temperature. By 3000 BC they were making glass beads for decoration, and by 2000 BC they were producing more complicated glass shapes, including bottles used mainly as containers for perfumes and other cosmetics.

Nowadays, a glass container such as a milk bottle or a light bulb is made by blowing a blob of molten glass into the required shape inside a metal mould. Ancient Egyptians made their bottles in a different way. They filled a cloth bag with sand, then dipped it into molten glass. The cloth burned away and the glass cooled around the sand to form the bottle.

CEMENTATION STEEL INVENTED c.1200 BC • **CHINESE CAST IRON MADE** c.400 BC

THE FIRST VEHICLES

In the late twentieth century we live in an age of transport. Personal cars, juggernaut lorries, high-speed trains, all run on wheels - and yet for most of their existence people lived without the wheel. Cave dwellers had no need of it, the first villagers got along quite well without it, and the Ancient Egyptians even built the pyramids with no help from wheels.

From where they quarried the giant stone building blocks, the Egyptians floated them down the River Nile on rafts, then pulled them on wooden rollers to the sites of the pyramids - an enormous task involving hundreds of thousands of labourers and slaves. Less heavy loads were transported on labourers' backs, or dragged along specially constructed roads on wooden sledges.

The Coming of the Wheel

Probably, the wheel was invented many times over at various times and places. People using long, thick wooden rollers to transport heavy objects must have noticed that a number of shorter, more lightweight rollers would do the same job, and when a roller was short enough it became a wheel.

By another mental jump, ancient people then added a wooden axle to link two solid wooden wheels centre to centre. If a sledge was now placed on top of the axle, the result was a sort of wheelbarrow. A sledge plus two pairs of wheels-and-axles made the first cart or wagon. But single slices of tree trunks were difficult to cut and likely to split. A further improvement was to make the wheels of several pieces of wood, braced together with struts. Wooden carts having this type of solid wheel are still to be seen, dragged by oxen or buffaloes, in many rural parts of the world.

Spokes and Chariots

The next improvement to the wheel was to make it more lightweight by hollowing it out. An added motive for this was probably war. Four-wheeled war chariots were invented by the Sumerians in about 2000 BC. These had wooden wheels hollowed out to make four spokes and a thick rim. The Assyrians who later ruled Mesopotamia were still more war-like. Their faster chariots, which were drawn by horses, had two wooden wheels with a much greater number of thinner spokes.

Above: Sumerian chariots with four solid wheels, dating from 4,600 years ago.

Rafts, Dugouts, and Canoes

To travel by river or across lakes or short stretches of sea, ancient people built rafts or canoes. Dugout canoes, still used on the giant rivers of Africa and South America, are simply hollowed-out tree trunks. Coracles are much smaller, basket-shaped canoes. Ancient Britons and others made theirs of plaited branches covered with waterproof skins or leather.

Ancient rafts sometimes made sea journeys of many thousands of kilometres. In modern times, the Norwegian explorer Thor Heyerdahl's *Kon-Tiki*, a Polynesian-type raft built of lightweight balsa wood logs lashed together with ropes, attempted to recreate one such journey. *Ra*, another of Heyerdahl's "prehistoric" boats, was constructed of river reeds. Small rafts and canoes were paddled or punted along. Larger canoes, large rafts like *Kon-Tiki* and reed boats like *Ra*, had a sail for propulsion by the wind. Still larger sea-going boats, which were used about 1200 BC by the first great sea traders, the Phoenicians, had both sails and many oars.

WHEEL INVENTED SEVERAL TIMES c.3200 BC • SPOKES INVENTED BY 2000 BC

Below: Nearly 5,000 years ago, Ancient Egyptians first built boats of papyrus reeds. These have long since decayed, but the sailing boat *Ra*, a modern version of the Egyptian boats, was built and sailed long distances by Thor Heyerdahl and his crew.

REED BOATS INVENTED c.2700 BC • OCEAN-GOING TRADING BOATS BY 1200 BC

GREEK INVENTIONS

We think of the Ancient Greeks as the inventors of science and philosophy. Of course, many good scientific brains existed long before the peak of Greek civilization in the sixth to third centuries BC. More than 2,000 years before this, Cheops of Egypt had built his great pyramid, 147 m high and covering 32 hectares, to an accuracy of about 1 cm!

However, it is true to think of the Greeks as having the first truly scientific civilization. In mathematics, they invented much of the algebra and geometry scientists and students still use today for their calculations. (Even though the most famous theorem in geometry, that called after the Greek Pythagoras, was actually invented 1,000 or more years earlier by an unknown Ancient Egyptian!)

Inventors with Names

It is during these Greek times that we begin to hear the names of actual inventors. Apart from Cheops and a very few others, they had previously been anonymous. In mathematics, Euclid the Greek (who lived about 300 BC) is the most famous of all geometers. Hero, a Greek who lived about 200 BC in the famous ancient "science city" of Alexandria on the mouth of the River Nile, invented a toy steam turbine; an instrument for cutting metal screws; and various devices using gear wheels. He is also credited with the invention of a heat-operated compressed air engine for opening temple doors.

The inventions of Ctesibius, an Alexandrian Greek who lived about a century later, include a water pump for fire-fighting, a water clock, a crossbow with pistons operated by compressed air, and a musical organ called the *hydraulis* which had pipes through which air was pumped with bellows. The bellows were inflated and deflated by flowing water.

These Greek mechanical inventions are ingenious but hardly of world-shaking practical importance. Some other known Greek inventors were more practical. Anaximander, in the fifth century BC, is said to have invented a map of the world. Glaucos of Chios (date unknown) was perhaps the first blacksmith to weld pieces of iron by hammering them together while red hot. Other, anonymous, Greek inventors greatly improved the wood-turning lathe, making possible more elegantly decorative

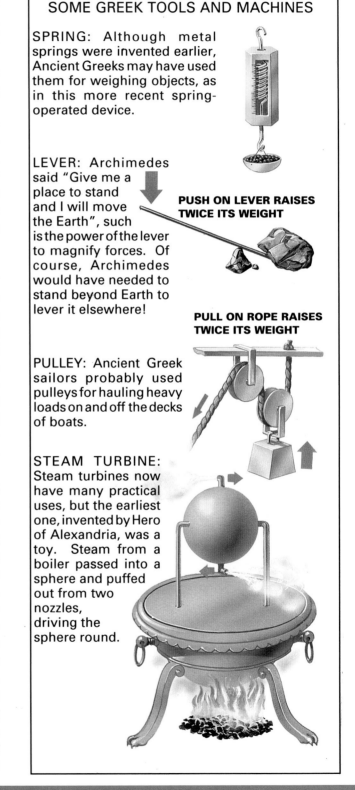

SOME GREEK TOOLS AND MACHINES

SPRING: Although metal springs were invented earlier, Ancient Greeks may have used them for weighing objects, as in this more recent spring-operated device.

LEVER: Archimedes said "Give me a place to stand and I will move the Earth", such is the power of the lever to magnify forces. Of course, Archimedes would have needed to stand beyond Earth to lever it elsewhere!

PUSH ON LEVER RAISES TWICE ITS WEIGHT

PULL ON ROPE RAISES TWICE ITS WEIGHT

PULLEY: Ancient Greek sailors probably used pulleys for hauling heavy loads on and off the decks of boats.

STEAM TURBINE: Steam turbines now have many practical uses, but the earliest one, invented by Hero of Alexandria, was a toy. Steam from a boiler passed into a sphere and puffed out from two nozzles, driving the sphere round.

chair legs and other wooden objects. The Pharos, a lighthouse in Alexandria, Egypt, was the greatest of its kind: over 76 m high, with an oil-lamp beam reaching 60 km out to sea.

In general, though, the Ancient Greeks seem to have been rather contemptuous of invention for merely practical purposes. They reserved their en-

thusiasm and brilliance for mathematics and other inventions of the mind alone. The ideas of Aristotle (384-322 BC) ruled scientific thought for the next 1,500 years. Archimedes (287-212 BC), who was a more practically-minded Greek scientist, created the science of mechanics on which most later working inventions have been based.

Below: Among the greatest building constructions of ancient times were amphitheatres used for staging plays and other shows. This one is in Dodona, Greece.

ROMAN INVENTIONS

The Roman Empire was the greatest the world had ever seen, extending west to Spain, north to Britain, south to north Africa, and east to the Caspian Sea in southern Russia. As a result of conquering all the countries that lay within this huge area, the Romans were influenced, to a greater or lesser degree, by their different cultures. By far the biggest influence came from Ancient Greece, the most civilized country in the Western world. The Romans borrowed Greek philosophy, Greek science, and Greek art. They even borrowed many Greek gods and goddesses to turn into their own!

The Romans were great conquerors but not, on the whole, original thinkers. Since inventing something demands original thought, neither were they great inventors. Their wealth and power allowed them to take advantage of the inventions of others such as the Greeks, and to make them bigger and better. Very often, the name of an inventor in the Roman Empire shows that he was in fact a Greek or other foreigner.

Inventions for War

Many inventions or improvements of Roman times have to do with making war. Large catapults and battering rams had been used for thousands of years to lay siege to fortified cities. The Romans made them even bigger. They often had fancy names: the scorpion was a catapult for firing spears; the falarica was a bent-back plank that when released, sent a missile flying towards the enemy. These were just two of many types of Roman siege weapons.

An even more effective war invention was the Roman army itself. Highly disciplined and efficient, its foot soldiers, armed with short swords and javelins, moved into attack behind the solid wall of their shields, usually driving a far less organized tribal enemy pell-mell before them.

Roads, Mills, Baths, and Aqueducts

The Ancient Romans are also famous for their dead-straight roads, built to speed their army through home and conquered territory. Although feats of civil engineering, these long slabstone stretches were actually rather clumsy and wasteful of labour. The same could certainly be said of Roman treadmills, on which countless thousands of slaves toiled out

their short lives, turning wheels to provide power for such tasks as pumping water, lifting building blocks, or grinding corn.

A more humane and inventive type of Roman mill was called after Vitruvius, although he had little to do with inventing it. Here, the wheel was turned by water, and the grindstone turned out anything up to 200 kg/h of flour. Earlier donkey mills had only been able to manage about 5 kg/h!

Vitruvius also described the drainage and sanitation installed in Roman cities. Again, this was nothing new: cities of the Indus Valley in India had had piped water 2,000 years earlier. But Roman city homes, with their hot-water baths, and central air heating provided by hypocausts (see right), certainly offered a new level of personal luxury. Many Roman aqueducts for carrying water into towns and cities are still standing today.

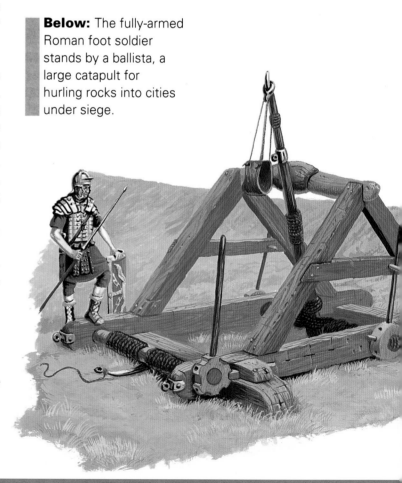

Below: The fully-armed Roman foot soldier stands by a ballista, a large catapult for hurling rocks into cities under siege.

Right: The houses of rich Roman citizens were often centrally heated with a hypocaust. This channelled hot air from a fire in a basement, through hollow tiles, to heat the rooms above.

HOT AIR TO UPPER ROOMS

FIRE

Left: The Roman aqueduct at Segovia, Spain was built 1,900 years ago, yet is still in good enough condition to carry water.

Below: This short stretch of Roman road is at the site of the ancient city of Carthage, in Tunisia.

CITY AQUEDUCTS FROM c.50 BC • BATHS AND CENTRAL HEATING c.50 BC

CHINESE INVENTIONS

China was by far the most inventive of ancient civilizations. It also long outlasted the Western civilizations of Greece and Rome. The age of Chinese inventions was an extremely long one, running from 2000 BC to late medieval times, when Marco Polo visited China and reported its wonders to the West.

Bronze, Ceramics, and Silk-Making

As early as 1500 BC, the Chinese melted bronze alloy and cast it into moulds to make bronze urns and other containers of a size and decorativeness unequalled elsewhere in the world. At the same time, their potters made beautiful and technically advanced pots of china clay (also known as kaolin).

Chinese luxury garments woven from silk date from as long ago as 1000 BC. This trade continues, basically unchanged, today. Silkworms, which are moth caterpillars, are put to feed on mulberry leaves, which are plentiful in China. The cocoons which the caterpillars spin are unwound to produce fine but strong silk threads. Each cocoon yields about 300 m of pale yellow silk.

Right: This young Chinese woman of about two centuries ago is spinning silk with a type of distaff and spindle.

SILK-MAKING c.1000 BC • PAPER-MAKING c.AD 100 • WOOD BLOCK PRINTING c.AD 800

26

Paper and Printing

The Chinese invented the art of paper-making in about AD 100 by mashing soft tree fibres in water, then pressing and drying the mass to make paper. Much later, in the eighth century AD, they invented printing, using wooden blocks with raised letters which were inked and pressed onto paper.

Clocks and Compasses

For thousands of years, both Chinese and Western civilizations measured the time of day by such primitive instruments as water clocks and shadow clocks or sundials. In AD 725, a Chinese inventor, Yi Hsing, described the first mechanical clock, the much more accurate type of clock we are familiar with today.

At about the same time, an unknown Chinese invented the magnetic compass. The first practical model took the form of a magnetized iron needle floating on a small straw mat on water. The needle always pointed in a north-south direction, to guide sailors and other travellers.

Below: In Chinese paper-making, fibrous pulp from trees was spread on a mesh. Water was squeezed out and the paper sheets shaken from the mesh.

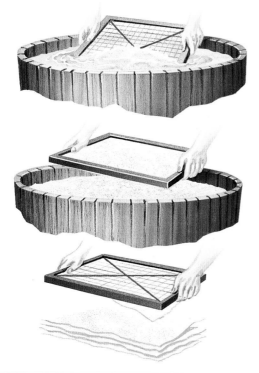

A Wealth of Inventions

Many other important Chinese inventions - and some not so important - are less easily dated. Perhaps as long as 2,000 years ago they invented the lateen sail for ships, though this triangular sail first appears on drawings of Greek ships. The stern-mounted rudder was another Chinese nautical invention. Also about this time, the Chinese invented the wheelbarrow!

About AD 500, an astronomical invention of the Chinese was the armillary sphere. This demonstrated the positions of the Earth and other planets compared with that of the Sun, and it shows that the Chinese had a much clearer idea of the Solar System than did the Western astronomers of the time. The Chinese also had a better knowledge of optics, since they invented optical lenses about AD 700.

Advances in Chinese medicine included herbals that described thousands of medical treatments, and surgical operations including that for removing cataracts from the eyes. Finally, the Chinese showed the extent of their civilization in about AD 800 by first inventing gunpowder, and then using it not for war but for fireworks!

Below: The earliest magnetic compass used a spoon made of an iron mineral called lodestone. The spoon's handle pointed north.

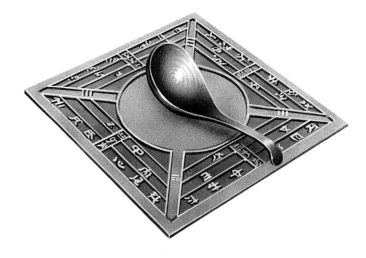

MEDIEVAL TECHNOLOGY

The final collapse of the Roman Empire in the fifth century AD was followed in the West by the Dark Ages. Constant warring between nations and tribes led to a state of confusion in which few new European inventions were made. As we saw on pages 26-27, Chinese civilization went on producing many important inventions, but the West only knew of these in the forms of silk and other art objects exchanged in trade.

Arab Science

In the sixth century AD, the Arabs founded their Middle Eastern Empire. They were not great inventors, but they became the custodians, or storers, of Greek science. In particular, they improved and developed mathematics and chemistry. *Algebra* is an Arabic word, and so is *alembic,* the name of a chemical apparatus for distilling, or purifying, liquids.

The Arabs borrowed some of their mathematics from India. It was in this country, about AD 500, that modern arithmetic began with the invention of a sign for zero. This new counting method of 1... 10... 100, etc., was far more efficient than the clumsy I, II, III... X... C, etc. of the Romans.

Science and Western Business

In the twelfth and thirteenth centuries AD, Arab and Indian mathematics and science were passed on to Western countries. The new, efficient number system, introduced into Europe by Leonardo Fibonacci, (AD 1180-1250), led to the invention of accountancy, so that better business records could be kept. This in turn led to the rapid expansion of trade and Western countries became far richer as a result. These changes marked the beginning of the Renaissance, or rebirth, of the West.

Cathedrals, Castles, and Technology

Already, in the medieval West, great things had been accomplished in building technology, as anyone who has visited a medieval, or Gothic, cathedral will know. The spires of these magnificent buildings, carrying their huge weight with the utmost delicacy, still soar over the cities of Europe. Their beautiful stained glass windows were another new technological invention of medieval times.

In Syria and other parts of the Middle East, other giant buildings of this time still rise starkly out of the desert. These are the Crusader castles, built as Christian strongholds from which European knights rode out to convert, and fight, the surrounding Muslim peoples.

Below: Making a cathedral window. One craftsman paints part of a design on a piece of glass. The design is baked hard in the firebowl. Then the stained glass pieces are set between strips of lead metal.

A more peaceful development in medieval building technology was the windmill. This was invented in Persia about AD 800, and by AD 1100 was widely used in Europe for milling flour. In European forests, in AD 1250 or thereabouts, another major leap forward in technology was the melting of iron using charcoal fanned white-hot in furnaces having the German name *Stücköfen*. For the first time in the West, iron could now be melted and cast into moulds to make many useful objects - 1,600 years after the Chinese had first invented cast iron!

Above: The beautiful fan-vaulting of King's College Chapel, Cambridge, dates from 1466-1515.

FIRST GOTHIC CATHEDRALS AD 1100 • IRON FIRST MELTED IN EUROPE C.AD 1250

THE AGE OF PRINTING

time, specialized learning became available to large numbers of people. From the new universities of Paris in France and Oxford in England, scholars passed on Greek science and philosophy, in the form of the new printed books, to anyone interested and rich enough. In cities and other business centres, banking was invented and printed paper money, or bank notes, began to be used for the first time.

Left: Johannes Gutenberg at his printing press, about 1450.

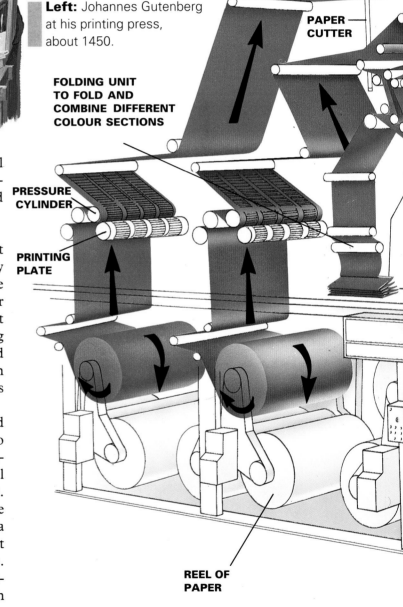

PAPER CUTTER

FOLDING UNIT TO FOLD AND COMBINE DIFFERENT COLOUR SECTIONS

PRESSURE CYLINDER

PRINTING PLATE

REEL OF PAPER

The Ancient Chinese, nearly always first in practical invention, began printing books using wooden printing blocks in the eighth century AD. Six hundred years later, printing was re-invented in the West.

Gutenberg's Bible

Johannes Gutenberg (1400-1468) printed the first Western book, his famous Bible, using a greatly improved method, with alphabetic letters of type metal. This is an alloy of lead metal with smaller amounts of tin and antimony. It melts easily but cools as a hard metal, so is very suitable for making letters for printing. A similar printing method had actually been invented at some time in the 1300s in far-off Korea, but Gutenberg's printing press was not influenced by this.

The metal letters (and numbers) were inked and used over and over, in different combinations, to print out different words and sentences. Each sentence was set up by the printer as a number of metal letters and blanks or spaces, called a type stick. Many type sticks were then set up together to make a whole page-worth of printing material, called a forme. Books hundreds of pages long could be set up and printed in hundreds or thousands of copies.

Like business arithmetic (see pages 28-29), printing was an enormously important invention which rapidly changed the Western world. For the first

METAL-TYPE PRINTING IN KOREA c.1350 • GUTENBERG'S BIBLE PRINTED 1448

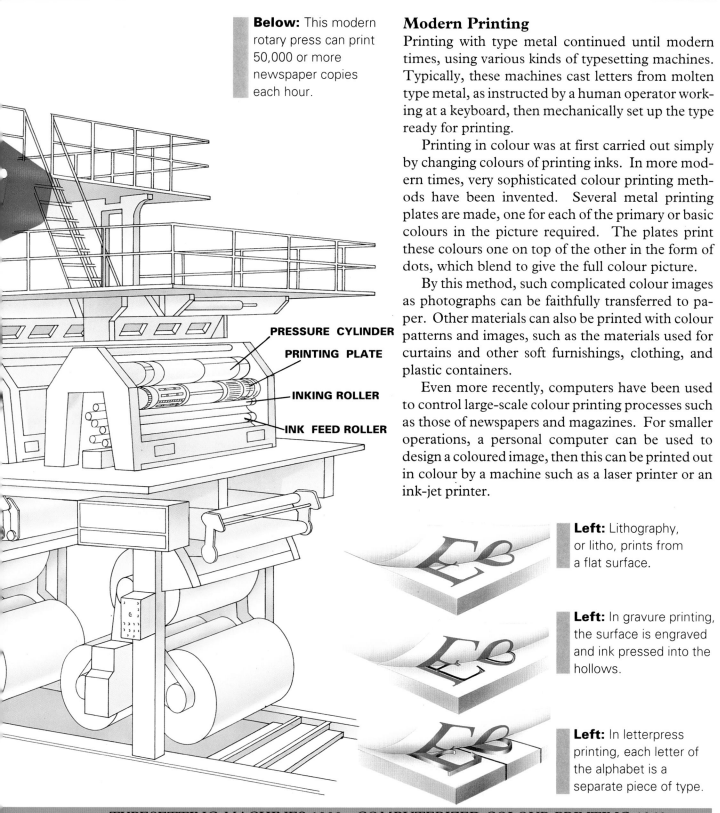

Below: This modern rotary press can print 50,000 or more newspaper copies each hour.

PRESSURE CYLINDER

PRINTING PLATE

INKING ROLLER

INK FEED ROLLER

Modern Printing

Printing with type metal continued until modern times, using various kinds of typesetting machines. Typically, these machines cast letters from molten type metal, as instructed by a human operator working at a keyboard, then mechanically set up the type ready for printing.

Printing in colour was at first carried out simply by changing colours of printing inks. In more modern times, very sophisticated colour printing methods have been invented. Several metal printing plates are made, one for each of the primary or basic colours in the picture required. The plates print these colours one on top of the other in the form of dots, which blend to give the full colour picture.

By this method, such complicated colour images as photographs can be faithfully transferred to paper. Other materials can also be printed with colour patterns and images, such as the materials used for curtains and other soft furnishings, clothing, and plastic containers.

Even more recently, computers have been used to control large-scale colour printing processes such as those of newspapers and magazines. For smaller operations, a personal computer can be used to design a coloured image, then this can be printed out in colour by a machine such as a laser printer or an ink-jet printer.

Left: Lithography, or litho, prints from a flat surface.

Left: In gravure printing, the surface is engraved and ink pressed into the hollows.

Left: In letterpress printing, each letter of the alphabet is a separate piece of type.

TYPESETTING MACHINES 1900 • COMPUTERIZED COLOUR PRINTING 1960

SHIPS AND NAVIGATION

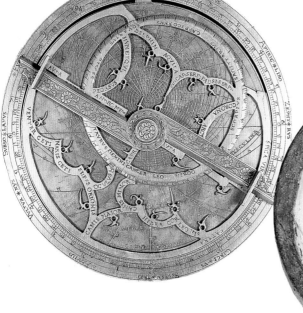

The Renaissance was the age when sailors first voyaged around the world. About 1460 a type of sailing ship suitable for such long sea voyages was built. Called the carrack, it was up to 500 tonnes in weight and had three or four masts. Larger versions, first built around AD 1550, were the galleons used by the Spanish Armada.

Bigger and better sea-going ships also benefited fishermen. Fishing trade increased enormously about 1450 with the invention of drift nets up to 110 m long, towed behind the ship to scoop up entire shoals of food fishes such as herring.

Latitude and Longitude

Long-distance voyagers continued to navigate their way by the positions of the Sun and stars. To measure their latitude, or distance north or south of the Equator, they invented such instruments as the cross-staff and quadrant, and later the sextant, which is still used today. These instruments of navigation all measure the angle between the Sun, or a star such as the Pole Star, and the horizon. A ship's navigator then checks this measurement on a chart showing how these angles vary through the year, and so obtains his latitude.

SEA-GOING CARRACK INVENTED c.1460 • MERCATOR'S MAP 1569 • GALLEON c.1550

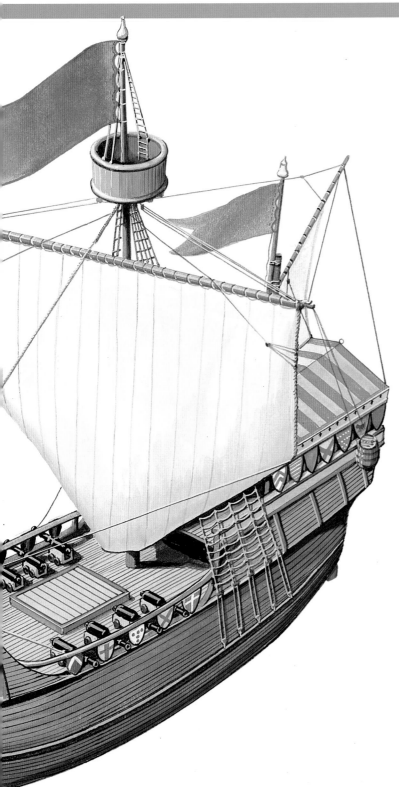

To find his longitude, or distance east or west of a line running from the North to the South Pole, a navigator measures how long it takes his ship to get from one point to another farther east or west. This time measurement, together with the ship's average speed, gives the distance travelled. The more accurately the time is measured, the more exactly the distance can be calculated, so the invention in 1761 of the chronometer, a very accurate ship's clock, by John Harrison (1693-1776), was a great advance in marine navigation.

Maps and Compasses

A navigator needs accurate maps on which to mark out, or plot, his course on the oceans of the world. Early maps of the world were often fanciful. Christopher Columbus (1451-1506) never knew that America, the country he so famously discovered, was not in Asia! A major problem for early map-makers was that the Earth is round (more or less), whereas a navigation map is flat. Gerardus Mercator (1512-1594) first solved this problem by projecting the shapes of the Earth's continents and oceans on to a flat surface. The Mercator Projection, as his invention is called, creates maps which show true directions or courses, although the precise distances still have to be worked out.

A navigator also needs a compass to show the exact direction in which his ship is travelling. The magnetic compass, invented by the Ancient Chinese (see pages 26-27) is still used by ocean navigators. Large ocean-going ships also have a more modern instrument called a gyrocompass. This is based on the principle that a gyroscope, or spinning top, resists any change in direction. When set to point due north, a gyrocompass will continue to do so, and unlike the magnetic compass it is unaffected by magnetic objects - such as other large steel ships passing close by!

Left: This carrack, dating from about 1460, was one of the first true ocean-going ships.

JOHN HARRISON'S CHRONOMETER 1761 • GYROSCOPIC SHIP'S COMPASS 1908

NEW INSTRUMENTS

Ideas and curiosity flowered in Renaissance Europe. This led, among a great variety of other things, to the invention of scientific instruments for measuring things in an exact way. Often, as we have seen before, the super-inventive Chinese had anticipated the Europeans by some hundreds of years, but it was in the West that the age of science finally began, around the year 1600.

Early scientific instruments were made with precision from various materials including wood, metals, and glass. Most favoured of the metals was brass, an alloy of copper and zinc which had been invented way back in Roman times. This alloy is fairly easy to form into special shapes, but at the same time is hard and permanent enough for precision instruments.

Glass and Lenses

Glass, too, had been invented in ancient times, by the Egyptians. Later, in about AD 700, the Chinese invented optical lenses by grinding and polishing clear glass into special shapes. Optical lenses refract, or change the direction of, light rays passing through them, in a precise way. The use of lenses that we are most familiar with is in the spectacles we wear to correct our sight.

The Ancient Chinese are not known to have used lenses in this way. A Christian monk, Roger Bacon (1214-1294), is often credited with inventing spectacles, though this is far from certain. In the early 1600s, at the beginning of the age of science, lenses were used for making the first microscopes and telescopes. These tremendously important inventions deserve more room for themselves, so are dealt with separately on pages 36-37.

Thermometers, Barometers, and Mirrors

One early scientific instrument using both brass and glass is the thermometer, for measuring temperature. Who first invented it, around 1600, is unknown. The glass tubes of the first thermometers contained alcohol, a liquid which expands as temperature rises, and contracts as temperature falls. Later thermometers, like the ones we use today, employed the liquid metal mercury.

Soon afterwards, in 1643, the barometer was invented by the Italian scientist Evangelista Torricelli

Below: The largest glass lenses of the eighteenth century were those of giant burning glasses, such as this French invention.

Right: A glass thermometer of the seventeenth century and a barometer of the early eighteenth century.

GLASS SPECTACLES INVENTED c.1250 • WEIGHT-DRIVEN IRON CLOCKS c.1350

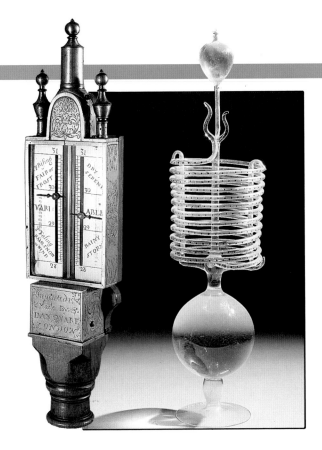

(1608-1647). A barometer measures the pressure of air, that is, of the Earth's atmosphere. This varies with the weather, so the barometer is useful for showing changes in weather.

From early times, glass was melted, then moulded or blown into shape, to make articles such as bottles. About 1600, in Italy, molten glass first began to be rolled out on metal tables where it cooled to form plate glass or glass sheet of an even thickness. This new technology led to the production of better quality glass windows. When the new glass sheet was coated on one side by a shiny metal, better mirrors resulted. Mercury and later silver were used in this way for backing mirrors.

Iron Clocks

Iron, meltable in the new furnaces, was at first too hard and tough a metal for making small precision instruments. Large instruments, however, could be made from it, the most famous of these being weight-driven clocks. A particularly well-known one is the clock of Salisbury Cathedral in Wiltshire. It dates from as early as 1386.

Below: Isaac Newton's glass prism splits sunlight into a coloured spectrum.

NEW WAYS OF SEEING

Below: Galileo made revolutionary discoveries with this early telescope.

EARLY TELESCOPES

Galileo's telescope of 1609, left, contained two lenses, a convex front lens and a concave eyepiece lens. This brought objects about thirty times "closer", but the image was blurred. In the next century or so an improved telescope was developed, of a type still popular today. It has a double or compound front lens, and a convex eyepiece lens. Objects viewed through it are magnified more and blurred less than with Galileo's telescope. The improved telescope also shows a larger area, or field of view. Its one defect is that it shows the viewed object upside down. However, this doesn't matter much if you are viewing the Moon, and not at all if you are viewing a distant star!

CONVEX EYEPIECE LENS

COMPOUND FRONT LENS

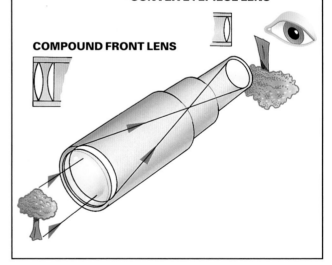

Movements of the Sun, Moon, planets and stars have fascinated people from the beginnings of human history. For a very long time, these mysterious heavenly bodies were associated with gods and magic. In the fifteenth century AD, the modern science of astronomy was born, when people first began to look at the heavens scientifically.

Telescopes and the Universe

A Dutchman called Hans Lippershey probably invented the first telescope in 1608. Within one year this was improved by one of the greatest of all scientists, the Italian Galileo Galilei (1564-1642). With his new telescope, Galileo made many astronomical discoveries. He saw, for example, that the giant planet Jupiter had four moons. Previously, only our own Earth was believed to have a moon.

These and other discoveries and experiments of Galileo finally upset the age-old picture of the Universe as Earth surrounded by Sun, Moon, planets, and stars. The new picture put the Sun, not Earth,

Below and right: With his powerful single-lens microscope, Anton van Leeuwenhoek examined details of tiny "animalcules" such as this water flea.

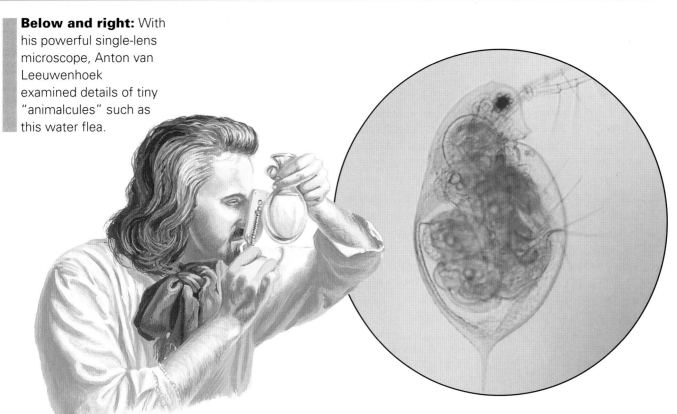

at the centre, with planets and their moons orbiting around it. Also, the new telescopes soon showed the stars to be enormously farther away than the other heavenly bodies.

Galileo's telescope is basically the type still used by amateur astronomers. It is a tube with a glass lens at either end. One section of the tube slides into the other to alter the distance between the lenses. By adjusting this distance, far-off objects are brought into focus and enlarged so as to seem closer.

Other types of telescope use mirrors as well as lenses. In this century, optical telescopes have grown extremely large, with a main glass mirror 7 m or more in diameter. Even greater in size are many radio telescopes. These giant metal dishes catch and focus not light waves, but radio waves reaching Earth, often from vastly remote parts of the Universe. The first radio telescope was built by an American, Karl Jansky, in 1931.

Microscopes

Another use for optical lenses is to magnify things so tiny that we cannot see them clearly, or at all, with the naked eye. The first powerful, accurate microscopes were invented by a Dutchman, Anton van Leeuwenhoek (1632-1723). His microscopes all had a single lens, but this was powerful enough to reveal a whole new, strange world of creatures and objects never seen before. Leeuwenhoek's "animalcules" included, for example, the single-cell forms of life we now call bacteria.

Optical microscopes became even more powerful with the addition of one or more extra accurate lenses. These compound microscopes can magnify objects up to a few thousand times, and are invaluable to scientists and medical doctors studying such things as microbes and details of the various cells of our bodies.

Electron microscopes, invented in the twentieth century, are vastly more powerful still. Their "lenses" are provided by strong magnets which bend and focus beams of electrons, as an optical glass lens does light rays. Magnifications of up to 1 million times reveal details, for example, of the smallest forms of life, the viruses.

SINGLE-LENS OPTICAL MICROSCOPE 1660 • ELECTRON MICROSCOPE 1932

IRON AND STEEL

Nowadays, anyone looking down from a bridge at a crowded motorway with thousands of motor vehicles speeding by, will get some idea of mass production. In ancient and medieval times, a few things such as salt and glass bottles were mass-produced. More complicated mass production had to wait for the Industrial Revolution to provide the necessary power and quantities of metals and other materials.

Coke and Blast Furnaces

The Industrial Revolution began in the early eighteenth century when coke, a fuel made from coal, was first used to make iron metal from its ores. This new process was called smelting. Abraham Darby (1677-1717) of Shropshire first used coke fuel to smelt iron ores in a blast furnace. Coke burns more efficiently and at a higher temperature than coal. As a result, Darby's coke-fired blast furnaces were able to produce iron metal in much larger quantities than had

Below: An iron foundry at the beginning of the nineteenth century.

Right: A modern blast furnace like this makes hundreds of tonnes of pig iron each day.

COKE AND IRON ORE

HOT AIR BLAST

MOLTEN SLAG

MOLTEN IRON

DARBY SMELTS IRON ORE WITH COKE 1708 • DAVY INVENTS MINERS' LAMP 1815

38

been possible in earlier, coal-fired blast furnaces.

With the greater amounts of iron available, much larger iron objects could be made. Abraham Darby's grandson, of the same name, built the world's first giant iron structure, a bridge, in 1779 (see page 43). Many complicated machines began to be made out of cast iron, including the steam engines we shall see on pages 40-41.

In England and other newly industrialized countries, coal mining increased greatly, providing more coke for the smoking blast furnaces. Coal mining was made safer in 1815 with the invention of the first miners' safety lamp by Sir Humphry Davy (1778-1829). The flame of this ingenious oil lamp burned inside a metal gauze shield, and would light a miner's way underground without causing explosions of mine gases.

The Age of Steel

Steel is an iron alloy containing a smaller and more precise quantity of carbon than cast iron. It is even tougher and stronger than cast iron, but was made in much smaller quantities until about 1870, when newly-invented types of blast furnace began to produce steel in batches of 25 tonnes or more.

One new steel furnace was the Bessemer Converter. It operated by blasting air not through the burning coke fuel as in Darby's blast furnace, but through the molten metal itself. This burned off most of the carbon in the metal, to make gases such as carbon monoxide. The gases bubbled up through the molten mass of metal and escaped from the furnace through a flue or chimney. By adjusting the air blast, a small but precise amount of carbon was left in the metal, just enough to make steel.

At the start of the twentieth century, everyone grew aware of the value of steel, as it became the new material for all sorts of machines, including motor cars. In 1908, Henry Ford in the USA invented mass production of motor cars with the manufacture of his Ford Model T family saloon.

Left: Stainless steel on a very large scale: the Thames Barrier in London, designed to prevent flooding.

MASS PRODUCTION OF STEEL FROM 1870 • MASS PRODUCTION OF CAR FROM 1908

STEAM POWER

Steam Pumps

In the Middle Ages, the power for turning a mill wheel was obtained from falling water, or from the wind, or from donkeys or other animals. Water-powered wheels were also used for pumping out flood water from coal and metal mine-workings. At the very end of the seventeenth century, many of these water-powered mine pumps were replaced by steam-powered pumps. Burning wood or coal to turn water into steam, these pumps were the earliest important industrial steam engines.

Thomas Savery (1650-1715) invented the first steam pump for mine drainage in 1698. A somewhat more efficient steam pump was invented by Thomas Newcomen (1663-1729) in 1705. These were both very large and clumsy machines. The pumping engine invented by a famous Scotsman, James Watt (1736-1819), in 1769, was much neater and more efficient. Another of Watt's important inventions was a governor, a device for controlling the speed of a steam engine. By 1782, Watt had designed further-improved steam engines of the sort to be used in only a few more years for the first steamboats and steam locomotives.

The Railway Age

Wagons running on rails, pulled by horses or pushed by men, had been employed since the 1500s for transporting coal and metal ores in mines. At first the rails were made of wood, then in the early 1700s, of iron. In 1804, the first steam locomotive to run on iron rails appeared. This was the invention of an Englishman, Richard Trevithick (1771-1833), who had previously built steam road carriages, which were less of a success.

Trevithick's steam locomotive of 1804 proved its worth by pulling five wagons filled with 10 tons of iron and 70 men a distance of five miles. It attained speeds of only a few km/h, but in 1829 the *Rocket* of George Stephenson (1781-1841) sped along on its rails at no less than 58 km/h. This famous little locomotive was of the type used on the first steam railway to carry both passengers and freight, the Liverpool and Manchester Railway of 1830.

Steamboats

In 1802, two years earlier even than Trevithick's first steam locomotive, the first successful steamboat came into operation. Named the *Charlotte Dundas*, she was powered by a new engine designed by James

NEWCOMEN'S STEAM PUMP

One of the earliest practical steam engines was the steam pump invented by Thomas Newcomen in 1705 to pump water out of flooded coal mines. It had a piston that was forced upwards by pressure of steam from a boiler. The steam was cooled and therefore condensed, creating a vacuum and allowing air pressure to force the piston down again. As the piston moved down, it dragged a pumping arm, which pulled water from the depths of the coal pit to the surface.

Right: From 1820 to 1870, the big Mississippi paddle steamboats dominated trade and social life in central USA.

FIRST STEAM PUMP 1698 • WATT'S STEAM ENGINE 1782 • STEPHENSON'S *ROCKET* 1829

Watt and constructed by William Symington of Glasgow, in Scotland.

The first steamboats, like many larger ones that later plied the waters of the Mississippi in the USA, were paddle steamers. The ship's screw propeller was invented in 1836 by an American, John Ericsson (1803-1889). From about 1850 onwards, most steam-powered boats and ships were propeller-driven. But often, they were built with masts and sails in case their steam engines broke down!

By the end of the nineteenth century, steamships included fighting ships and ocean liners built of steel. The fuel for their steam engines continued to be coal until the early 1900s, when oil fuel began to replace it. At the same time, a more efficient kind of steam engine, the steam turbine, began to be used in ships. Invented by an Ancient Greek more than 2,000 years before (see page 22), it was only now proving really useful!

FIRST STEAMBOAT 1802 • SHIP'S PROPELLER 1836 • LARGE STEAM TURBINES BY 1900

ROADS AND BRIDGES

Below: Motorways and other modern roads are built following exact engineering principles.

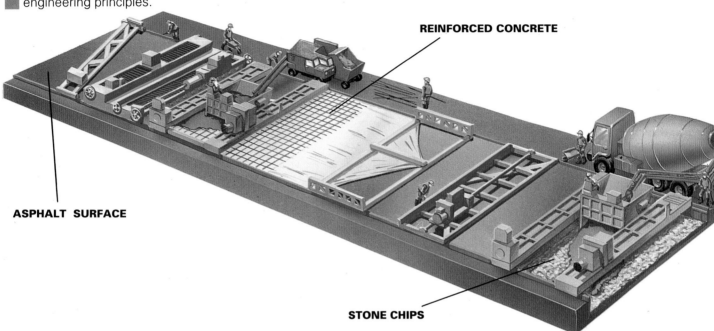

REINFORCED CONCRETE

ASPHALT SURFACE

STONE CHIPS

Roads

Among ancient road-makers, the Romans are famous for their arrow-straight roads. Roman road-makers first levelled the ground. Next, they laid down foundations of sand mixed with gravel or lime. Finally, they wedged stone slabs into this mixture to make the road surface. Sometimes, they cambered the road, that is they sloped it downwards at the edges. This caused rainwater to drain away so that the road did not become waterlogged.

This method of road-making sounds sophisticated, but crudely-slabbed Roman roads would soon ruin your shoes or bicycle wheels! In fact, roads remained crude constructions for a very long time. Even main roads of the nineteenth century in the most advanced countries often remained unpaved, and were muddy mires in wet weather.

Late in the nineteenth century, roads began to be paved with tarmacadam, a mixture of small stones bound together with tar, or the tarry substances asphalt or bitumen. This tarry surface dries hard, wears well, and is waterproof. Modern motorways and other major roads are constructed of reinforced concrete, with waterproofing layers either on top or below. The road may be electrically heated to melt any ice on its surface.

Tunnels and Canals

The longest tunnel of ancient times was 5.6 km long, dug through Mount Salviano in Italy with iron picks by Roman slaves. Tunnels of the nineteenth century reached 13 km long and were dug for railways. In 1861 pneumatic drills, operated by compressed air, were first employed for tunnelling, after explosives had been used to remove larger amounts of rock.

Modern tunnels, such as the "Chunnel" linking England and France, are cut with huge tunnelling machines. The earliest of these, called a tunnelling shield, was invented in 1818 by the famous British engineer Isambard Kingdom Brunel (1806-1859). His railway tunnels were lined with brick, whereas modern tunnels use reinforced concrete.

The earliest canals, or man-made waterways, were dug out by the Ancient Mesopotamians 5,000 years ago. About AD 100, the Chinese invented the

FIRST ALL-METAL BRIDGE 1779 • PNEUMATIC DRILL 1861• ALPINE TUNNEL 1871

first locks to connect stretches of a canal at different levels. This allowed canals, and boats on them, to travel across sloping ground. Famous modern ship canals include the Suez Canal, opened in 1869, and the Panama Canal, opened in 1914.

Bridges

Wooden, stone, and rope bridges date from ancient times. The first all-metal bridge still spans the River Severn at Ironbridge, Shropshire. Erected in 1779, it is built entirely of iron made in the first blast furnaces of the Industrial Revolution.

Most metal bridges date from much later in the twentieth century. They are built from steel, together with reinforced concrete, which itself contains strengthening steel bars. The various types of large bridge include beam bridges, arch bridges, suspension bridges, and cantilever bridges. One of the longest bridges in the world is the Lake Ponchartrain Causeway in Louisiana, USA, which extends for more than 38 km.

Below: A giant tunnelling machine cutting the Channel Tunnel or "Chunnel".

Above: Abraham Darby's iron bridge over the River Severn, which was built in 1779.

TEXTILES

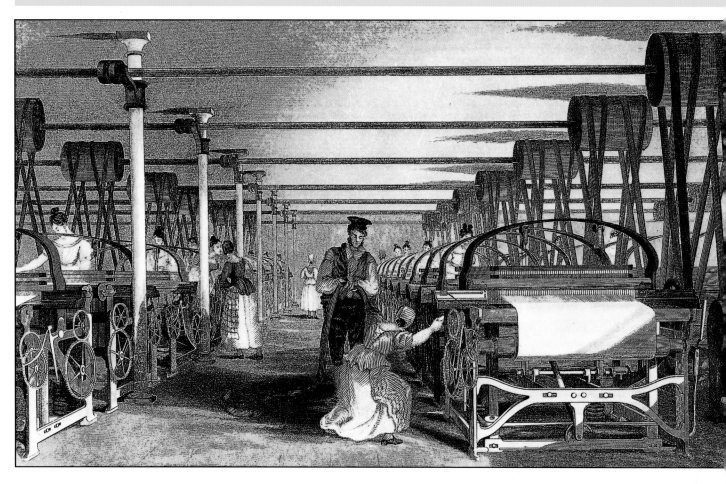

Above: A cotton-weaving
factory in England, 1835.

Spinning, Weaving, and Knitting

Clothes-making is almost as old as humankind. In
the traditional process, textiles for clothes, carpets,
and hangings are woven from plant fibres such as
cotton, or animal fibres such as wool. First, these
fibres have to be drawn out and twisted to make yarn
or thread. Then, cloth is woven using a loom, which
interweaves the threads.

In very early times, threads were spun with a
simple hand-held wooden instrument called a distaff
and spindle. In the early 1300s the spinning wheel
appeared, which speeded this process up. The spinning wheel was the ancestor of many later spinning
machines. These included the spinning jenny, invented by James Hargreaves in 1764, the spinning

SPINNING WHEEL INVENTED c.1300 • SPINNING JENNY 1764 • SPINNING MULE 1779

frame, invented by Richard Arkwright in 1769, and the spinning mule, invented by Samuel Crompton in 1779. These machines spun several yarns at the same time.

The loom was improved by the invention, by John Kay in 1733, of the flying shuttle, which speeded up the interweaving process. Steam-powered looms date from 1787 when one was invented by Edmund Cartwright. All these British inventions had the effect of creating the first large cotton cloth industry, which was based in Lancashire. From the early nineteenth century, cloth-making from cotton took place in factory rooms containing rows of spinning machines and mechanized looms.

Knitting is another way of making cloth and garments. The first knitting machine, built around 1600, was called, for an obvious reason, the stocking frame. The first factory-operated knitting machines were used for lace-making in Nottingham in the early 1800s.

Dyes

Clothes and other textiles are nearly always dyed in suitable colours before being sold in the shops. Early cloth dyes were mostly coloured substances obtained from plants, such as indigo (blue), madder (red), and saffron (yellow).

Artificial or man-made dyes came into use much later. The first artificial dye was invented in the USA by William Perkin in 1856. Called mauvein, from its colour, it was made from aniline, a chemical product of the coal tar industry. This and the later petroleum industry have given us most of our modern dyes.

Synthetic Textiles

The label on an item of clothing often tells us that the garment is made fully, or partly, of a synthetic fibre such as nylon or terylene. These and other synthetic fibres are products of the plastics industry, which grew in size and production after World War II. Plastics are manufactured chemically by a process called polymerization, which combines small chemical molecules together to make giant molecules, or polymers. From the polymers, synthetic fibres are obtained by a spinning process rather like that used by a spider.

Synthetic fibres such as nylon are much stronger

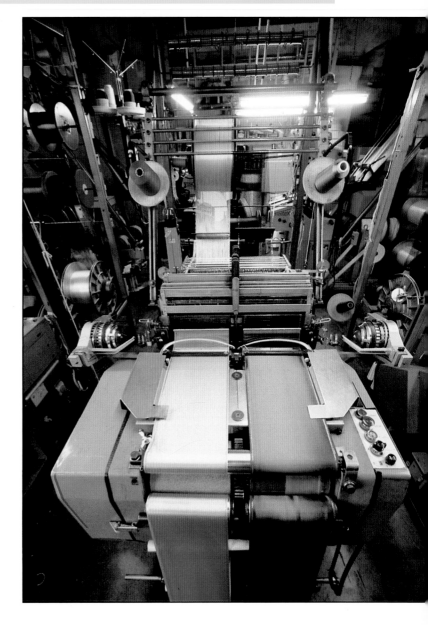

Above: Modern textile machines produce natural and synthetic fabrics at high speed.
Far left: Hand-dyeing cloth with indigo in Nigeria.

than natural plant fibres, and synthetic garments are harder-wearing - though many people still prefer cotton to nylon! Like natural fibres, synthetic fibres can be dyed a wide range of colours.

THE NEW AGRICULTURE

Ploughing and Sowing

Medieval farming proceeded as it had for a thousand years past, with small farmers using primitive tools. One of the few great improvements was the use of a mould board on the plough. This turned over the ploughed soil more efficiently, so aerating the soil more and increasing its fertility. The mould board was invented by the Ancient Chinese but was not used by Western farmers before AD 1000.

Seed was scattered or sowed into ploughed furrows by hand. Early in the seventeenth century, seeding machines began to be used for planting seeds automatically. The best known of these is the seed drill invented in about 1701 by Jethro Tull (1674-1741) in Oxfordshire. This and later seeding machines planted a quantity of manure or other fertilizer at the same time as the seed, in order to help plants grow.

Jethro Tull is also credited with the invention of an early harrow for rooting up weeds and loosening soil. Harrows began to be used widely in the early nineteenth century, and can still be seen in country places today. Typically, they have a heavy metal frame with spikes or sharp-edged discs for disturb-ing the soil, and are dragged along either by a horse or a tractor.

Harvesting and Threshing

Farmers went on reaping their wheat or barley with sickles and scythes up until the late eighteenth century, when the first reaping machines were invented. In England, Henry Ogle's reaping machine of 1822

Right: A threshing machine of the 1850s. It is driven by a steam engine.

TULL INVENTS SEED DRILL 1701 • OGLE INVENTS REAPING MACHINE 1822

had a knife-blade to cut the crop and a reel of revolving bars to push the cut crop on to a platform.

To thresh grain, or beat it out from its husk, farmers had for countless centuries used a hand-held flail (see pages 8-9). The first threshing machine, which also featured a sort of flail, was invented by a Scotsman, Andrew Meikle, in 1786.

Combine harvesters, or combines, are so called because they combine both reaping and threshing. In addition, they automatically bind up stalks into bundles for haystacks or other forms of storage. The first combine harvester was invented by an American, Samuel Lane, in 1828.

These first combines were drawn behind horses. Later, larger, combines were steam-powered. Modern combines include giant diesel- or petrol-engined machines such as those that harvest the wheatlands of the USA. Special machines have also been invented for harvesting other crops, including cotton, potatoes, sugar cane, sugar beet, peanuts, and various fruits.

Intensive Farming

Since World War II, advanced methods of farming have led to intensive culture of plants and animals. Single crops of wheat are grown on a huge scale, with much use of artificial fertilizers and pesticides. In battery farming, tens of thousands of chickens may be kept in a single shed and fed automatically. Cows may never leave their indoor stalls, where they are fed and milked by automatic machinery.

THRESHING MACHINE INVENTED 1786 • COMBINE HARVESTER 1828

ELECTRICITY

We owe our modern world of technology and industry mainly to steel and electricity. Both of these began to be made in large quantities in the late nineteenth century. The development of electric power came about after vital discoveries earlier in the century. The most important of these occurred when, in 1831, the great British scientist Michael

Faraday (1791-1867) invented a machine for making electric current - the first electric generator.

Electric Power Stations

From Faraday's small laboratory invention arose the giant electric generators of today's power stations. The first power station to provide electricity to houses and street lamps was the work of Thomas

Below: Thomas Edison in his laboratory. In front of him is one of his many inventions, an electric pen.

Edison (1847-1931), the great American inventor. In 1882 his electric power station started operation in the Pearl Street District of New York, USA.

Power stations feature another invention of Michael Faraday's, the electric transformer. This is a device for altering electric voltage and current from one level to another. In power stations, transformers raise the voltage of electricity supplied by generators to the very high voltage needed for distributing the electricity over long distances on the pylons and cables of the electric grid.

Electric Motors and Lamps

Faraday also invented the electric motor, a sort of electric generator in reverse. Electric current is supplied to the motor, which then supplies mechanical power to wheels or other moving parts. Electric motors are used to power a host of machines, from electric shavers to railway locomotives.

Another revolutionary invention of Thomas Edison, in 1879, was the incandescent filament lamp, or light bulb, which brought electric light into homes all over the world.

Electric Cells

Electricity can also be created by chemical reactions. Any device which stores chemical energy and delivers electric current is called an electric cell or battery. An early electric battery was invented in 1792 by the Italian Alessandro Volta (1745-1827). Modern chemical batteries include the familiar car battery or accumulator, and smaller dry batteries for electric torches.

Fuel cells are a twentieth-century invention for providing electricity. They are devices that use a continuous supply of fuel gases, which react together chemically. The energy of these chemical reactions is then produced as electric current. The uses of fuel cells include power packs in spacecraft.

Electric solar cells are another modern invention. They obtain their power from the Sun and convert this light energy directly into electricity. These cells are used in space research, and even more recently, large batteries of solar cells are becoming important for heating and lighting houses and a great variety of other buildings.

NON-POLLUTING POWER

Most electric power stations are polluting, but non-polluting power can be obtained in many ways:
1 By damming a lake or river, then releasing water through turbines, which generate electricity. **2** By damming a river estuary, then allowing water from the tides to flow through turbines. **3** By putting many small water turbines in the sea, each of which makes electricity from the waves. **4** By causing the wind to turn giant windmills or electric generators. **5** By using the power of sunlight, for example by converting its heat and light into electricity with banks of silicon cells. **6** By using heat in the Earth's crust to boil water to make steam. This is then used to operate steam turbo-generators to make electricity.

1 HYDROELECTRIC POWER

2 TIDAL POWER

HIGH TIDE

3 WAVE POWER

DAM

LOW TIDE

WAVE

TURBINE

4 WIND POWER

5 SOLAR POWER

BANKS OF SOLAR CELLS

6 GEOTHERMAL POWER

EDISON'S ELECTRIC LIGHT BULB 1879 • EDISON'S ELECTRIC POWER STATION 1882

TELECOMMUNICATIONS

To communicate with one another at considerable distances, people once had to rely on letters carried by land and sea. With the tremendous increase in trade and travel during the Industrial Revolution, businessmen and others needed to know more quickly what was happening, both at home and abroad. By the mid-nineteenth century, new discoveries and inventions in electricity (see pages 48-49) made this faster communication possible.

Telegraph and Morse Code
In 1837 in the USA, Samuel Morse invented the first telegraph, an instrument for sending electrical messages from one place to another through a wire cable. He also invented a code of long and short pulses of electricity, or "dashes and dots", known as the Morse Code, which was used to send these messages. Alfred Vail, who worked with Morse, invented a hand-operated metal contact lever or key for sending Morse-coded messages along the telegraph cable. In the 1850s, Morse's first long-distance electric telegraph began to operate over the 65 km distance between Baltimore and Washington in the USA.

Later telegraph cables were much longer than this, some extending for thousands of kilometres under the Atlantic and Pacific Oceans. Eventually, electric telegraph messages could be sent right round the world. To send these messages more efficiently, special typewriters called teleprinters were invented in the 1930s. Messages typed out at one end were almost immediately printed out automatically at the other. By this time, long-distance messages could also be transmitted without the need for a cable, by radio telegraphy (see pages 64-65).

Telephone
A telegraph message along a cable needs to be in code, but a telephone allows people to talk to each other directly. The telephone was invented in the USA in 1876 by Alexander Graham Bell, a Scots-born American. A year later, his invention was

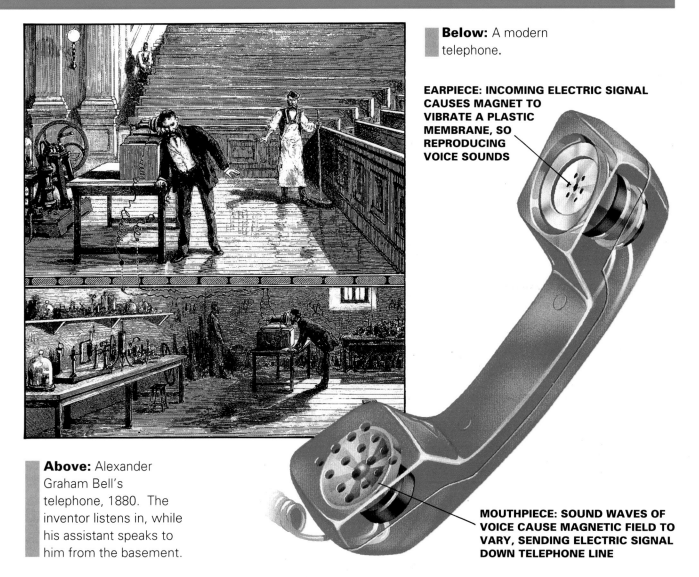

Below: A modern telephone.

EARPIECE: INCOMING ELECTRIC SIGNAL CAUSES MAGNET TO VIBRATE A PLASTIC MEMBRANE, SO REPRODUCING VOICE SOUNDS

Above: Alexander Graham Bell's telephone, 1880. The inventor listens in, while his assistant speaks to him from the basement.

MOUTHPIECE: SOUND WAVES OF VOICE CAUSE MAGNETIC FIELD TO VARY, SENDING ELECTRIC SIGNAL DOWN TELEPHONE LINE

improved by the American inventor, Thomas Edison.

Edison's improved telephone, like the telephones we use today, contained another new invention, the microphone. This has a diaphragm, rather like an eardrum, and an electromagnet connected to the telephone wire. Someone speaking into the microphone causes the diaphragm to vibrate, which in turn causes variations in an electric current flowing through the electromagnet. At the other end of the telephone wire, this varying electric current is turned back into sound, so reproducing the voice message.

The video telephone is a really new invention, using which people making and receiving a call can see, as well as hear, one another. The video telephone uses fibre optics (see pages 72-73) to carry the visual image.

Telemetry

By the 1930s, scientists were sending balloons high up into the Earth's atmosphere to record the weather. The weather balloons sent back their scientific data automatically, in the form of radio messages. Other man-made objects in the sky, such as aircraft, could be remote-controlled by radio signals. This is called telemetry. Examples of telemetry today include the control and relay of automatic messages from space satellites and probes.

PHOTOGRAPHY

Before the nineteenth century, images of people and objects were mostly made by artists such as painters and sculptors. Another early means of creating an image, known since the time of the Ancient Greeks, was the camera obscura. This is basically a darkened room or box ("camera" means "room" in Latin) with a small hole in one wall to let in daylight. An image of the scene outside then appears on the opposite wall. Artists later made use of the camera obscura for such tasks as drawing outlines of objects and scenes in correct perspective.

Camera and Photograph

The camera obscura is the ancestor of the modern photographic camera, which is also basically a box with a hole in it. Fitted into the hole is a transparent lens to let in light. The light from an object falls on to a light-sensitive film, which then records the object's image, called a photograph.

Photography as a whole was not invented by any one person - many people made photographic discoveries and inventions. In 1725 a German chemist, Johann Schulze, showed that the chemical compound silver nitrate turned black when exposed to light. In 1816 a Frenchman, Nicéphore Niépce, obtained a blackish photographic image using paper immersed in a solution of silver chloride.

Niépce could not "fix" this image to make it stay on the paper. He achieved this later, in 1826, together with another Frenchman, Jacques Daguerre. The first satisfactory photographs, appearing in 1829, were called daguerreotypes after this last inventor. Daguerreotypes were positive photographic images, fixed directly on to a copper plate using sodium thiosulphate, a chemical compound still used in photographic fixers.

For most modern photographs, a negative image is first made, from which many positives can be reproduced. The inventor of the first type of negative-positive photography, in 1839, was an Englishman, William Fox Talbot. However, it took various experimenters another 50 years to produce the flexible rolls of photographic film we use today.

Black-and-White and Colour

In modern photography, a roll of photographic film, contained in a light-proof cassette, is first placed in

Below: In a modern reflex camera, light enters the lens and is reflected by a mirror and a prism into a viewfinder, showing the photographer the scene. A light-exposure meter tells the photographer what shutter speed and lens aperture (opening) he should use.

Below right: William Fox Talbot of England made this early box camera in the 1840s. He fitted it with a lens from his microscope and pinned light-sensitive paper onto the inside of the hinged back of the camera.

SHUTTER

EXPOSURE METER

MIRROR

FILM

a camera. For black-and-white photography, the film, made of a plastics material, is coated with a solution of silver bromide. When light is allowed into the camera by clicking the shutter, a negative or reversed black-and-white image is formed on the film. When the film is removed from the camera, a positive black-and-white image is produced on paper by developing and printing the film using various chemicals. This positive print is the photograph.

In colour photography, the photographic film contains chemical dyes sensitive to coloured light.

The first modern colour negative films were made in the 1930s. Like black-and-white negatives, these films need to be developed and printed to make the final photograph. Modern colour slides or transparencies were introduced later, in 1946. They only need to be developed to produce the final coloured photographic image.

The Polaroid camera has a special colour film that contains both dyes and developer, which are released when the shot is taken, producing a colour photograph, which is then pulled from the camera.

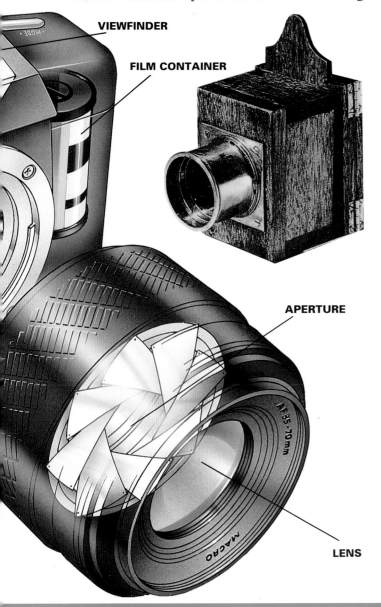

VIEWFINDER

FILM CONTAINER

APERTURE

LENS

MOVING PICTURES

The cinema shows us many photographs that follow one another so quickly that we see them not as separate pictures but as a continuous scene or action. The first cine-camera was invented by Etienne-Jules Marey in 1882. It was a gun camera that was first sighted on the object to be photographed, then took 12 photographs each second when the trigger was pulled. In more modern types of cine-camera, photographic film from a reel is fed through the camera while a shutter lets in and cuts off light at short intervals. The series of photographic negatives taken in this way is "played back" with a cinema projector, as the motion picture or film that we watch.

FIRST PHOTOGRAPHIC NEGATIVE 1839 • FIRST ROLL FILM 1885 • COLOUR FILM 1930s

Electricity was first supplied to light people's homes in the early 1900s. It brought with it the possibility of many other domestic inventions. Most of these were labour-saving devices. Others, such as radios and television sets (see pages 64-65) were more like luxuries. Labour-saving inventions gave people more time for listening and looking.

Cleaning

To remove dust from our homes we use a vacuum cleaner. A typical vacuum cleaner first brushes dust out of a carpet, then sucks it up with a suction fan rotated by an electric motor. Other fittings allow the vacuum cleaner to be used for cleaning curtains, cushions, and walls. The first vacuum cleaner was invented in 1901 by an American, Hubert Booth.

To clean our clothes we use a washing machine, also operated by an electric motor. Items to be washed go into a tub, together with a cupful of detergent or washing powder for loosening the dirt. In older-type machines, the motor turns or vibrates an agitator inside the tub to move the items around and wash the dirt out of them. In more recent washing machines, the motor spins the tub containing the washing. Washed clothes are dried either in a separate, electrically-heated drier, or automatically in the washing tub itself. The first electrically-powered washing machines date from just after World War I, but fully automatic machines were not manufactured and sold before 1937.

Sewing Machine and Safety Pin

Many families own a sewing machine for mending and making clothes. The first successful domestic sewing machine was invented in 1844 by Elias Howe in the USA. It was an improvement on an earlier machine of Walter Hunt, who, in 1849, also invented the safety pin.

Cooking and Storing Food

For cooking food we mainly use cookers heated by electricity or gas. Since the 1970s, microwave ovens have become popular. These cook food much more quickly because they heat up not the metal of the stove or pot but water in the food itself. The microwave was invented in the USA in 1945 by Percy Spenser for cooking popcorn.

For keeping food cool and fresh we use refrigerators and freezers. Nineteenth-century freezers were merely ice-boxes, but from the mid-twentieth century onwards, most household refrigerators had an electric motor. This pumps a cooling gas or liquid through pipes around the food.

Air Conditioning Homes

In countries that are cold some or all of the time, a modern house will be heated throughout by central heating. This can take the form of underfloor electric heaters, or heated air or water can be circulated through pipes and radiators. Heat for the air or water is provided by burning coal, oil, or gas in an electrically-controlled burner.

In hot countries a house may need to be cooled, and this is achieved with an air cooler, which refrigerates the air passing through it. Houses in countries with both hot and cold seasons may have a combined air-conditioning system.

Right: A modern house contains many gadgets for saving energy and making housework easier.

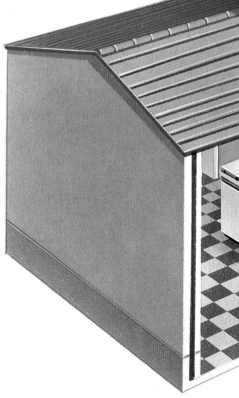

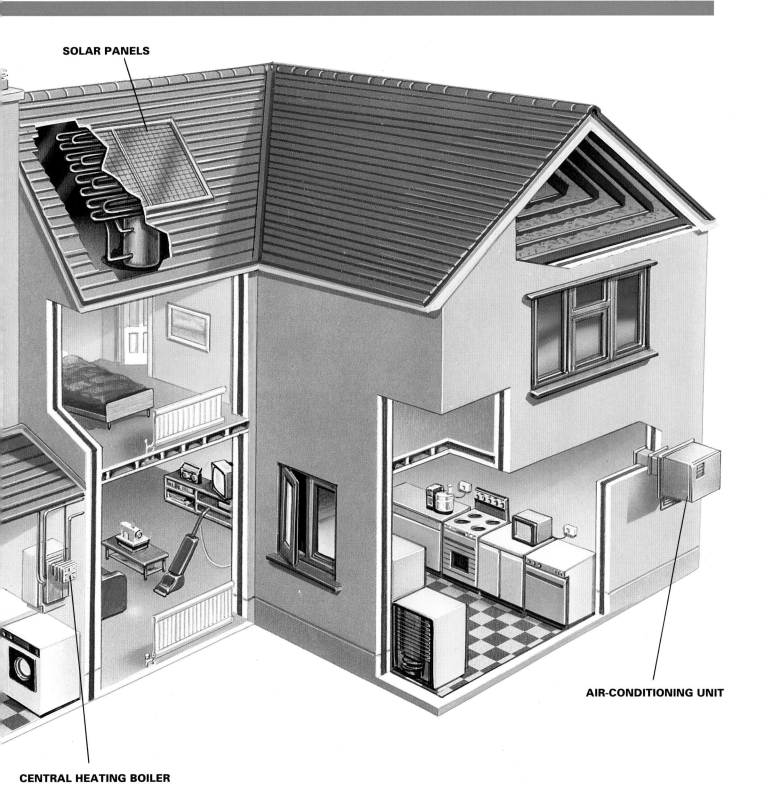

SOLAR PANELS

CENTRAL HEATING BOILER

AIR-CONDITIONING UNIT

CENTRAL HEATING INVENTED 1930s • MICROWAVE OVEN 1945

MODERN MEDICINE

Immunization

Perhaps the earliest invention of modern medicine was vaccination against the disease smallpox. This was first done as long ago as 1796, by an Englishman, Edward Jenner (1749-1823). The next form of vaccination came nearly 90 years later, when the great French scientist Louis Pasteur (1822-1895) vaccinated a boy against the terrible virus disease rabies in 1885. Thanks to these pioneers, we have vaccines today to protect or immunize us against many infectious diseases.

Surgical Operations

Operations were made far less painful by the invention of anaesthetics. In 1844 an American dentist, Horace Wells, first used nitrous oxide, or "laughing gas" as it was called, to make a patient less conscious of pain. At about the same time, the gas ether was used as a general anaesthetic to "put patients out" by two American doctors, Crawford W. Long and W.T.G. Morton.

In 1847 another famous early anaesthetic, chloroform, was first employed by a Scottish surgeon, James Simpson. In today's operations, these early general anaesthetics have been replaced by others which are less poisonous. Another safety invention for operations was the use of an antiseptic by an English surgeon, Joseph Lister (1827-1912). In the 1870s he used a spray of the antiseptic substance phenol, to kill any microbes that might infect surgical wounds. Lister's invention was the start of modern asepsis or microbe-free operating rooms.

Drugs against Disease

The earliest modern man-made drug was aspirin, which was first used in 1893 to relieve headaches and rheumatic pains. In 1910 the drug Salvarsan was invented by a German scientist, Paul Ehrlich (1854-1915). This was also known as "the magic bullet" because it was the first drug specially designed to cure a particular disease, syphilis.

In 1932, the first of a whole range of powerful

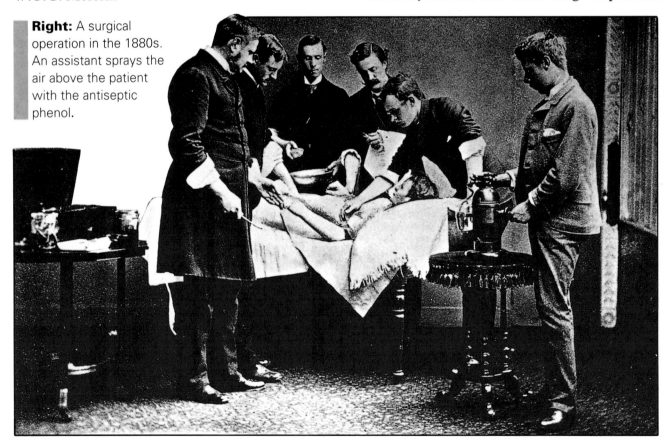

Right: A surgical operation in the 1880s. An assistant sprays the air above the patient with the antiseptic phenol.

microbe-killing drugs, the sulfa drugs, was invented by a German scientist, Gerhard Domagk. Sulfa drugs revolutionized the treatment of infectious disease, but were succeeded after World War II by the much less poisonous drugs called antibiotics.

Most famous of antibiotics is penicillin. This was discovered in 1928 by a Scotsman, Alexander Fleming (1881-1955), but was first manufactured and used to treat sufferers in 1941 by two other scientists working in Britain, Howard Florey and Ernst Chain.

Recent medical drugs include those designed to prevent a patient rejecting a new heart or other transplanted organ. Yet more new drugs have been invented to control mental illness, allowing patients to live normal lives outside mental hospitals.

Scanning Machines

X-rays, which penetrate solids such as the human body, were discovered in 1895 by the Dutch scientist Wilhelm Röntgen (1845-1923). In the early years of the twentieth century, the first X-ray ma-

Right: This modern X-ray machine scans a patient's brain to detect disease and other abnormalities.
Inset: A picture of the patient's brain appears on the monitor screen.

chines were invented for the examination, or scanning, of bones and other internal parts of the body.

More recent body scanners include ultrasound machines, which send out high-frequency sound waves and are used to examine babies still inside their mothers' wombs. Still other scanners are the ECG machine, which records details of heartbeat, and the EEG machine, which records electrical waves given out by the brain.

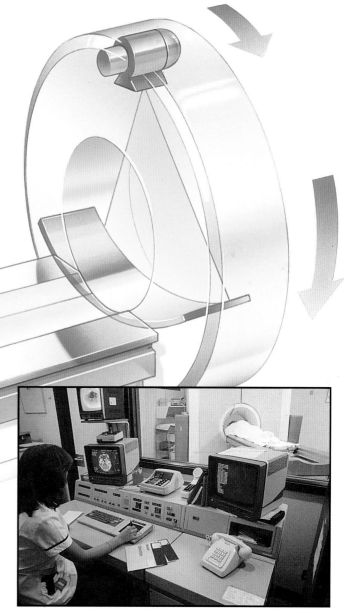

ROAD VEHICLES

The vast majority of cars, vans, and lorries we see on our crowded roads are powered by petrol or diesel engines. These are internal combustion engines, which get their power from explosions of fuel vapour or gas inside closed metal cylinders. The power of the explosions forces pistons up and down in the cylinders. This motion is transmitted, first to turn a crankshaft, and eventually to turn the vehicle's wheels. The ancestor of most modern internal combustion engines was invented in 1876 by a German engineer, Nikolaus Otto.

Another type of internal combustion engine was invented in 1956 by the German engineer Karl Wankel. The Wankel engine has no pistons. Instead, fuel vapour explosions cause a rotor to move around inside the engine casing. As in the Otto engine, this movement is then transmitted to the vehicle's wheels.

Motor Cars

The very first motor cars were also German inventions. They had petrol engines and were designed and built by Karl Benz and Gottlieb Daimler in 1885. Daimler and Benz soon became famous names in the world of motor cars.

Not long afterwards, the first diesel-engined cars appeared on the roads. The diesel engine was invented by Rudolf Diesel in Germany in 1897. In petrol engines, a mixture of fuel gas and air is exploded inside each cylinder by sparks from a spark plug. In diesel engines the gas-air mixture is exploded by pressure alone.

The first motor cars were luxury vehicles, built by hand in small workshops. For ordinary people to be able to afford them, motor cars had to be built more cheaply and in larger numbers. This was first achieved by Henry Ford in the USA in 1908. He invented the process whereby motor cars are assembled by many workers on a mass-production line. The Model T Ford, most famous of family cars, was mass-produced for 19 years until 1927.

Bicycles and Motorcycles

First ancestor of the bicycle was the "hobby horse", a crude two-wheeled vehicle "walked along" by young men-about-town, about the year 1815. The first

Below: A cutaway view of a motor car engine, showing how it burns petrol fuel to provide power.

Below: Road trains like this one are commonly seen thundering through the Australian outback.

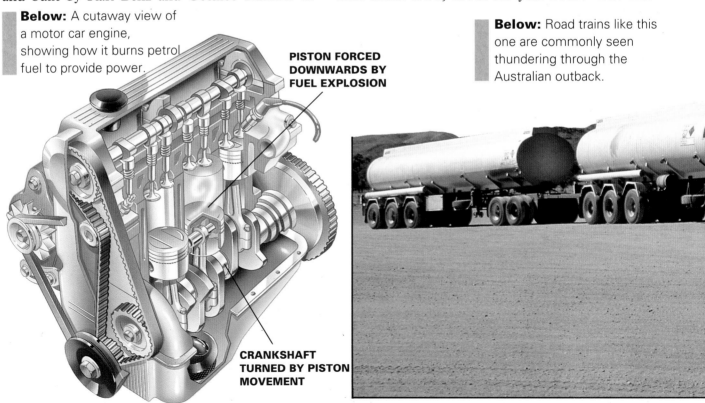

PISTON FORCED DOWNWARDS BY FUEL EXPLOSION

CRANKSHAFT TURNED BY PISTON MOVEMENT

BICYCLE FROM 1815 • MOTORCYCLE FROM 1870 • ELECTRIC TRAM FROM c.1880

pedal-driven bicycle was invented in 1839 by a Scotsman, Kirkpatrick Macmillan. Inflatable rubber tyres were added later, having been invented in 1889 by a Northern Irishman, John Dunlop.

The first motorcycles, invented in France by Pierre and Ernest Michaux about 1870, were dangerously hot, steam-powered machines. Petrol-engined motorcycles were much safer. Earliest of these was a motor tricycle invented in 1884 by an Englishman, Edward Butler. A year later, a petrol-engined motor bicycle was built by the German inventor Gottlieb Daimler.

Electric Road Vehicles

Slow-moving milk floats are familiar electric vehicles on our roads, and small electric cars are becoming more popular as city runabouts. Such road vehicles are driven by an electric motor powered by a battery. Electric trams and buses are driven by engines powered from an overhead line. The electric road carriage was invented as long ago as 1837, by Robert Davidson, a Scottish engineer.

ELECTRIC CARS

Small electric cars like this French one are beginning to appear in our cities. They travel at reasonable speeds and, unlike most other road vehicles, do not pollute the atmosphere. Their power comes from an efficient electric battery, which at present can be used for a maximum of about 100 km, after which the owner must recharge the battery by plugging it into the electricity mains supply overnight.

AIRCRAFT

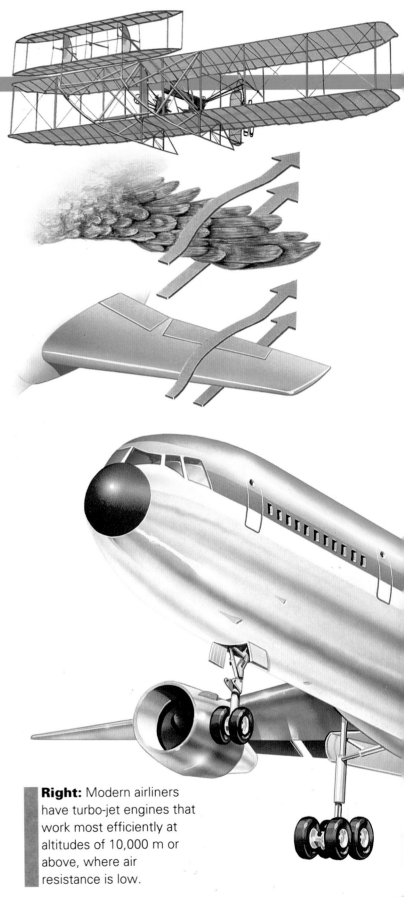

Balloons

Men first flew through the air when, in 1783, the Montgolfier brothers used a hot-air balloon to fly over Paris. Hot air weighs less than colder air, so that a balloon filled with hot air will rise up in the Earth's atmosphere. The balloon will then be driven along by the wind.

The Montgolfier brothers heated the air in their balloon with a fire in a bowl suspended beneath the open bottom end of the balloon. Hot-air ballooning is again popular today. Air in modern balloons is heated with a propane fuel gas burner supplied from a gas bottle.

In the early part of this century, balloons filled with hydrogen gas were more often seen. These included giant airships, called zeppelins, which bombed London in World War I. Hydrogen gas is much lighter than air, but it is also dangerously flammable. After a number of terrible disasters in which airships caught fire, a safer gas, helium, was used. Helium is the lightest gas after hydrogen, but it is very expensive. Because hydrogen was dangerous and helium expensive, giant airships soon became a thing of the past, though large gas-filled balloons are still used for such purposes as advertising.

Aeroplanes

An aeroplane is an aircraft that has wings or other structures that provide lift. When air passes across an aeroplane wing, the air pressure below the wing is higher than the pressure above the wing. This pressure difference increases as the aeroplane speeds up on the runway. When the pressure difference exceeds the weight of the aeroplane, the plane is lifted into the air, or takes off.

The earliest aeroplanes were model gliders. One built by Sir George Cayley (1773-1853) flew in England in 1804. In 1849 Cayley built a larger triplane, or three-winged glider, which carried a boy, the first person to fly in an aeroplane, a distance of a few metres.

The USA saw the real beginnings of modern aeroplane flight when in 1903, the brothers Orville and Wilbur Wright flew their petrol-engined biplane *Flyer 1* a distance of 260 m. Petrol-engined aeroplanes rapidly increased in size, speed, and safety.

Right: Modern airliners have turbo-jet engines that work most efficiently at altitudes of 10,000 m or above, where air resistance is low.

HOT-AIR BALLOON FLIGHT 1783 • WRIGHT BROTHERS' POWERED FLIGHT 1903

60

HOVERCRAFT

Hovercraft are really aircraft because they travel over water or land on a cushion of air. The air cushion is provided by turbines, which also turn propellers to push the hovercraft along.

Left: Air passing over and under the wings of an aircraft and a bird. The moving air provides lift, which keeps both plane and bird aloft.

Top left: The Wright *Flyer*, the first successful powered aeroplane.

Above: Helicopter blades act like revolving wings to provide lift, as shown in this French-made *Ecureuil*.

In 1909 Louis Blériot flew his monoplane from France to England across the English Channel. From large bombers used in World War I were developed the first airliners to carry passengers from country to country. Biggest of all were passenger seaplanes that took off and landed on water. The German Dornier Do-X seaplane of 1930 had 12 engines and carried 150 passengers at 400 km/h.

Jet Aeroplanes

The fastest and biggest aeroplanes today are powered by jet engines. Air enters the front of the engine and is used to burn fuel injected into the engine. The hot gases produced by this burning expand and blast

out from the back of the engine, so thrusting the plane forward. In the 1930s, Frank Whittle in England developed the gas turbine jet engine that powered warplanes as well as post-war jets, but the first jet aeroplane of all to fly, in 1939, was the German Heinkel He 178.

BLERIOT FLIES ENGLISH CHANNEL 1909 • HEINKEL HE 178 JET PLANE FLIES 1939

SOUND RECORDING

Left: A piano recording being made in France in the 1880s. Thomas Edison's phonograph and some cumbersome "horns" are being used.

AMPLIFIER

MICROPHONE

SOUND

SOUND

We hear sounds by means of air vibrations called sound waves which, in turn, cause our eardrums to vibrate. This is also the principle used to make gramophone records. An eardrum-like diaphragm is made to vibrate by sound waves. These vibrations cause a sharp stylus to cut grooves into a disc of plastics material. Sound can then be played back from the grooves, using another hard stylus, or needle. To be audible, the recorded sounds must be made louder, or amplified.

The inventor of sound recording, in 1877, was Thomas Edison (see also pages 48-51). His phonograph recorded sound as a pattern of little dents on a sheet of tinfoil. Ten years later, also in the USA, Emile Berliner invented a method of recording sound on rotating, spirally-grooved discs - the first gramophone records. Early recordings were made loud enough for people to hear by means of a trumpet-shaped amplifier.

Tape Recording

In 1898 a Danish engineer, Valdemar Poulsen, invented a method of recording an electrical signal in the form of patterns of magnetism on a metal wire. This method became the basis for magnetic sound recording, familiar to us as tape recording.

Right: Making and playing an audio recording. The violin's sounds are picked up by the microphone and recorded on a cassette, which is played back through a loudspeaker.

EDISON INVENTS PHONOGRAPH 1877 • INVENTION OF RADIO VALVE 1907

Recording tapes are the work of many inventors, including J.A. O'Neil in the USA, who in 1927 invented a paper tape coated with a magnetic material. Modern tapes are made of flexible plastics coated with a material containing magnetic substances such as oxides of iron and chromium.

Tape recordings are made using a microphone (also invented by Thomas Edison) which converts sound waves into electrical signals. These signals then cause magnetic patterns to form on a magnetic tape. When the tape is played back on a tape machine, pick-up heads convert the magnetic pattern back into an electrical signal. The signal is amplified, then converted into sound waves.

The first modern electrical amplifier was the vacuum tube, or radio valve, invented in 1907 by Lee de Forest in the USA. The amplified electrical signal is fed into a loudspeaker, which contains an electromagnet. The signal causes this to vibrate, so producing sound waves.

Compact Disc Recording

In the early 1980s a new type of sound record began to be sold. The compact disc or CD is made of a plastics material, inside which is a spiral of very small "pits", lying at different depths. These have been cut out with a laser beam (see pages 72-73) in a recording studio. In a CD player, the compact disc is scanned by another laser beam which reflects from the non-pitted parts of the disc. The laser reflections are "read" by a light sensor which turns them into electrical signals. These are turned back into sound as in other types of sound or audio equipment. The great advantage of the compact disc is that the laser beam which is used for scanning causes absolutely no wear and tear, unlike a record player needle, or even the playing head of a tape player.

CDs are now also available for video recording of television programmes (see pages 64-65). In this process, both vision and sound are recorded on a compact disc.

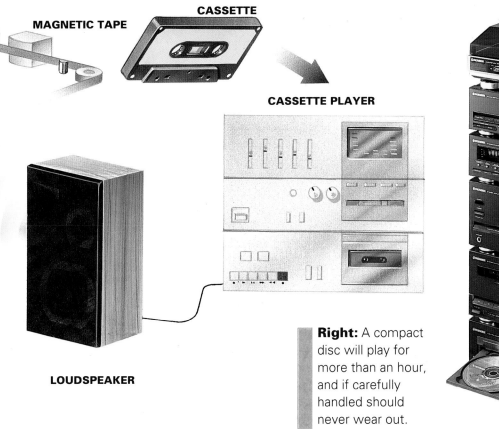

MAGNETIC TAPE

CASSETTE

CASSETTE PLAYER

LOUDSPEAKER

Right: A compact disc will play for more than an hour, and if carefully handled should never wear out.

FIRST ELECTRICAL RECORDING 1898 • FIRST TAPE RECORDING 1920s

RADIO AND TELEVISION

Radio

Radio waves were discovered in 1887 by Heinrich Hertz in Germany. He detected these waves with a metal wire loop in which the radio waves caused an electric current to flow. Hertz's wire loop was, in fact, the first radio receiving aerial or antenna.

The inventor of radio communication was an Italian, Guglielmo Marconi (1874-1937). In 1895, he used electric sparks to make radio waves travel outwards from an aerial. This was the first radio transmitter. Also, using a radio receiver like that discovered by Hertz, he detected his own radio wave signal at a distance of more than 1 km.

Marconi transmitted his radio messages in Morse Code (see pages 50-51). This was called wireless (radio) telegraphy - since, unlike earlier telegraphy, it needed no wire cable. By 1901 Marconi could transmit radio messages all the way across the Atlantic, and wireless telegraphy became important for passing code messages to and from ships at sea.

Even so, Marconi's early electric spark transmitter was a very inefficient way of making radio waves. The next leap ahead came with the invention in 1907 of the vacuum tube or radio valve (see pages 62-63). In an improved type of transmitter, radio

Above: Guglielmo Marconi, inventor of radio broadcasting.

valves amplified electrical signals to make more powerful radio waves.

Radio receivers were also greatly improved so that for the first time, human speech and musical sounds could be heard clearly. By the early 1920s, public radio broadcasting had begun, and people were beginning to listen in at home with their own "wireless sets".

Television

This form of public broadcasting transmits TV signals in the form of radio waves, which are received by the aerials of our TV sets, and turned into pictures on our screens. The first successful TV transmissions, in 1925, were those of a Scottish inventor, John Logie Baird. However, Baird's TV system was soon supplanted by another system based on a TV camera invented in the same year by Vladimir Zworykin in the USA.

In a TV studio, the scene to be broadcast is filmed with this special type of camera, which converts light signals into electrical signals. The signals are then

MARCONI INVENTS RADIO BROADCASTING 1895 • FIRST HOME RADIOS 1920s

amplified and transmitted as radio waves. At the same time, sound for the TV programme is recorded and also transmitted as radio waves. These waves are received by our TV aerials and turned back into sound and pictures in our TV sets.

A TV set contains a cathode ray tube, invented in 1897 by Karl Braun in Germany. The wider end of the cathode ray tube is the TV screen. Its narrower end, inside the TV set, contains a "gun" that fires a beam of electric particles, or electrons, at the screen. As they hit the screen, the electrons cause a special chemical coating, called a phosphor, to glow brightly.

The electron gun rapidly scans the TV screen from top to bottom. If there is no TV programme being received, an overall glow appears on the screen. An electrical TV signal modifies the intensity of the electron beam so that a picture appears on the screen.

Left: The inside of a colour TV set seen from the back. The red, blue, and green signals which combine to make the picture are directed onto the screen by three separate electron guns.

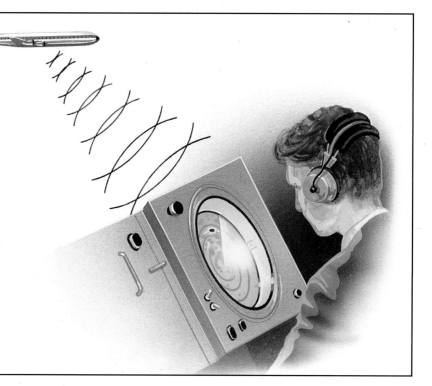

RADAR

The word RADAR stands for "RAdio Detection And Ranging". This means that radar can detect an object at a distance from us and also tell us how far away the object is. It does both these things by means of radio waves, which travel out from a radar aerial to hit the object and reflect back from it, to be detected by the aerial. The time taken by the waves to travel out and reflect back measures the distance of the object. An image of the object also appears on a radar screen, similar to a TV screen. Radar was invented in 1935 by Robert Watson-Watt, and used for defence against enemy aircraft in the Battle of Britain. Since that time, radar has also been used widely on ships and civilian aircraft.

CATHODE RAY TUBE INVENTED 1897 • ZWORYKIN'S TV CAMERA INVENTED 1925

OIL AND GAS

Fossil Fuels

Like coal, petroleum oil and natural gas are fossil fuels. Coal was formed in the Earth's crust from trees and other plants that lived hundreds of millions of years ago. Similarly, oil and gas were formed in the Earth's crust from the decayed bodies of countless tiny creatures that once flourished near the surface of seas. All fossil fuels are known chemically as hydrocarbons because they consist largely of the chemical elements hydrogen and carbon. When they are burned in air, fossil fuels release both heat and the greenhouse gas, carbon dioxide.

Oil and Gas Prospecting

Petroleum oil is obtained mainly by drilling into the Earth's crust. The first oil well was drilled in Pennsylvania, USA, by Edwin Drake in 1859. To find out where oil lay, prospectors or explorers at first merely looked for where it seeped up to the Earth's surface.

At the beginning of the twentieth century, they began to look for much larger and deeper oil reserves by methods of geophysical prospecting. In one widely-used method, prospectors set off explosions on the surface, which send sound waves down into the Earth's crust. By examining the sound reflections from these explosions, geophysicists can tell where and at what depth oil lies.

By 1918, oil wells as deep as 6,000 m had been drilled. Modern drilling bits or cutters, at the end of the drilling stems or pipes, are tipped with extra-hard materials such as tungsten carbide or small diamonds, which cut through even the hardest rock. When oil is reached deep down, it is often under great pressure and pumps itself up to the surface.

Below: A dirty town gasworks in the mid-nineteenth century.

DRAKE'S FIRST OIL DRILLING RIG 1859 • OIL WELLS DRILLED 6,000 M DEEP BY 1918

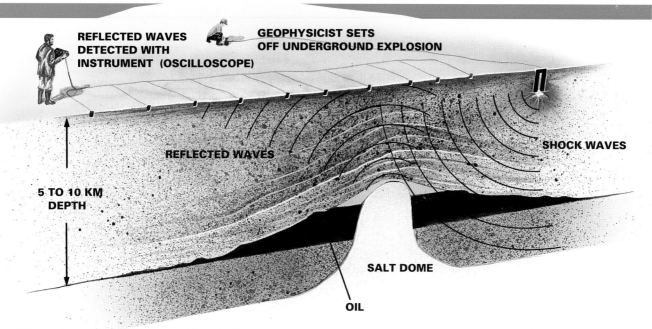

REFLECTED WAVES DETECTED WITH INSTRUMENT (OSCILLOSCOPE)

GEOPHYSICIST SETS OFF UNDERGROUND EXPLOSION

REFLECTED WAVES

SHOCK WAVES

5 TO 10 KM DEPTH

SALT DOME

OIL

Natural gas has been formed by similar processes to petroleum oil, and so is often found in the same places, under land or sea. For undersea oil or gas, huge drilling platforms are floated out to sea and anchored on the sea bed. Drilling then takes place in a similar way to land drilling. Natural gas consists mostly of the gas methane. It is piped away in huge volumes from the gas wells, both to industry and for burning in the cookers and gas heaters of our homes.

Petroleum Refining

Oil is pumped away from drilling rigs to be transported and refined. After World War II, bigger and bigger ships were built for transporting crude oil. A supertanker of today carries up to half a million tonnes. Crude oil is transported and pumped to an oil refinery, where it is heated and distilled to make a range of useful petroleum products. These distillation products may then be used more or less directly, as fuel oils for vehicle engines, such as petrol and diesel oil for road and rail vehicles, and as aviation fuel for aircraft. Similar refinery products include lubricating oils for reducing wear by friction in engines and other machines.

Other products of petroleum distillation are supplied to the chemical industry as raw materials. These are used to make a tremendous range of plastics, explosives, medical drugs, pesticides, synthetic fabrics, and pharmaceuticals.

Above: A simplified diagram of one geophysical method of prospecting for underground oil.

Below: An oil drilling platform off the Australian coast.

NUCLEAR POWER

All solids, liquids, and gases are made up of one or more types of chemical element, and 92 different elements occur in nature. Atoms are the smallest parts into which any chemical element can be divided or split up. A simple example is the element neon, the gas used in luminous signs, which consists solely of a mass of whirling neon atoms.

Splitting the Atom

The central part of an atom is called the nucleus, and this, too, can be split. The first scientist to split an atomic nucleus, in 1919, was an Englishman, Ernest Rutherford (1871-1937). He bombarded the gas nitrogen with sub-atomic "bullets" called alpha particles. They split the nuclei of many nitrogen atoms, turning these into atoms of another gas, oxygen. This process of splitting atomic nuclei is called nuclear fission.

The nuclei that Rutherford split also released energy in the form of very fast-moving sub-atomic particles, called protons. This is one example of nuclear energy. Another example is the nuclear radiation given off by atomic nuclei which spontaneously split up. Chemical elements having this unstable type of atom are said to be radioactive. They have many uses in science and technology.

Nuclear Reactors

Rutherford's discovery showed that atomic nuclei could release large amounts of energy. Scientists

Below: A cutaway view of a nuclear reactor used in the electricity generating industry.

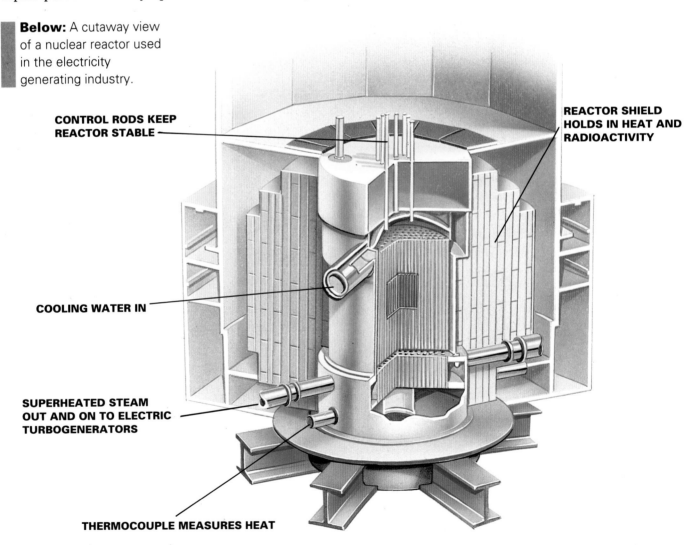

CONTROL RODS KEEP REACTOR STABLE

REACTOR SHIELD HOLDS IN HEAT AND RADIOACTIVITY

COOLING WATER IN

SUPERHEATED STEAM OUT AND ON TO ELECTRIC TURBOGENERATORS

THERMOCOUPLE MEASURES HEAT

RUTHERFORD SPLITS ATOM 1919 • FERMI BUILDS NUCLEAR REACTOR 1942

then began to ask how this energy could be produced and controlled on a large scale, to provide nuclear power. The answer came in 1942, when the Italian scientist Enrico Fermi (1901-1954) led a team in the building of the world's first nuclear reactor at the University of Chicago in the USA.

Nuclear reactors are the power units of nuclear power stations, which nowadays make a proportion of the world's electricity, and also provide nuclear explosives for weapons. In a nuclear reactor, energy is released from radioactive fuels in a controlled way. Heat made by the reactor is used to turn water into steam, which is then used to power steam turbogenerators that make electricity.

Below: Looking down into the cooling water of another type of nuclear reactor, the LIDO, at Harwell, England.

Thermonuclear Power

The Sun and other stars get their vast energy by another nuclear process, the combining together of atomic nuclei at temperatures of hundreds of millions of degrees. This thermal, or heated, combining of nuclei is called nuclear fusion, and the energy it produces, thermonuclear energy.

When nuclei of hydrogen atoms combine at these super-high temperatures, they form nuclei of atoms of the next heaviest gas, helium. However, some matter is left over, which turns into energy. As Albert Einstein (1879-1955) showed by his famous equation $e = mc^2$, even a tiny bit of matter turns into an enormous amount of energy.

Thermonuclear energy might provide all the power the world needs, if only it could be controlled. Inventions such as the *Tokamak* in the USSR, and *JET* in Europe are attempts to do this, but so far the only examples of man-made thermonuclear power are the devastating explosions of H-bombs.

Below: The *JET* experimental thermonuclear reactor.

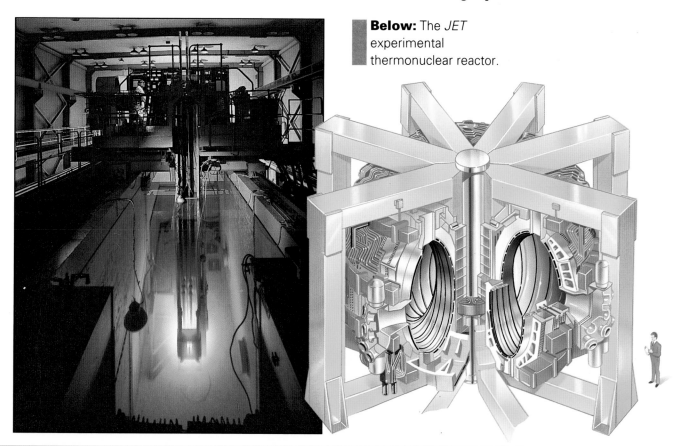

NUCLEAR POWER STATION 1954 • H-BOMBS 1950s • *TOKAMAK* AND *JET* 1960s

CHEMICAL INDUSTRY

Since the time of the Ancient Egyptians, people have made and used chemical substances. For thousands of years they have extracted salt from saltpans dried out in the sun, made glass from sand, and manufactured paints, pigments, and cosmetics - to give but a few examples of early chemical industry.

This chemistry, however, was always of a very practical kind. Useful substances were made without the scientific knowledge of just *how* they were made. The science of chemistry was often mixed up with the superstitions of alchemy. True chemical science had to wait until around 1766, when an Englishman, Henry Cavendish (1731-1810), made the first scientific description of the simplest of chemical substances, the gas hydrogen.

Large-Scale Chemistry

The earliest example of modern chemical industry was a large-scale process for making soda, or sodium carbonate, invented in 1787 by Nicolas Leblanc (1742-1806). The Leblanc process needed large volumes of another major chemical, sulphuric acid. This was quickly supplied by the lead chamber process, which was invented by a Scottish chemist, John Roebuck (1718-1794).

With the aid of these "bulk" chemicals, several more industrial chemicals were soon being made, including alkalis for large-scale soap manufacture, and hydrochloric acid, which was a by-product of this process. High explosives were another innovation of the mid-nineteenth century, with the discovery by A. Sobrero in 1846 of the violently explosive compound nitroglycerine, which was then used in 1866 by Alfred Nobel (1833-1896) to invent a much safer high explosive, dynamite.

Twentieth-Century Chemistry

A major chemical invention of the early twentieth century was the Haber process for making ammonia from hydrogen gas, together with nitrogen gas from the air. This process, invented in 1908 by a German, Fritz Haber (1868-1934), used an iron catalyst (a catalyst is a substance which speeds up chemical activity without itself changing) to make the gases combine. Both ammonia and metal catalysts became highly important for the next revolutionary chemical invention, plastics.

The first plastics material to be invented, in 1909, was bakelite, named after the Belgian chemist Leo Baekeland (1863-1944). Like other plastics, it is made by a polymerization process in which simple chemical substances, called monomers, are made to combine together to form very complex chemical substances called polymers or plastics.

Rubber is another important type of polymer. Since the early nineteenth century, natural rubber has been obtained from trees, but in 1910 a synthetic rubber was invented by the Russian, Sergei Lebedev (1874-1943). Both natural and synthetic rubber are used, for example, in motor car tyres.

Organic chemicals is the name given to an enormous number of chemical products, including dyes, inks, paints, solvents such as benzene, monomers for plastics, pesticides, perfumes, sweeteners such as saccharine, and medicines such as aspirin. These chemicals first began to be manufactured from coal tar in the mid-nineteenth century, and began to play a major part in industry in the twentieth century.

Below and right: This sulphuric acid plant burns the yellow sulphur in a furnace and converts the gases to acid with the aid of a metal catalyst.

LEBLANC PROCESS INVENTED 1787 • DYNAMITE 1866 • HABER PROCESS 1908

70

INDUSTRIAL CATALYSTS 1908 • FIRST PLASTICS 1909 • SYNTHETIC RUBBER 1910

SCIENTIFIC INSTRUMENTS

CHROMATOGRAPHY

Chromatography is a method used to discover what a chemical mixture is made of. In paper chromatography, the mixture is dissolved in a suitable liquid. Then a spot of this is placed near the lower edge of a strip of porous paper. The paper is hung so that this edge is immersed in a solvent liquid. The solvent rises up the paper, carrying the substances of the mixture with it, but the different substances rise at different rates, so that they separate out. Then, by various means, the identity and amount of each substance is calculated. Column chromatography works on the same principle, but instead of porous paper, uses a column of porous material in which the various substances in a mixture are made to separate out.

Chemical Analysis

Finding out what is in chemical substances, and how much, is called chemical analysis. One important method for analysing mixtures is chromatography, invented in 1903 by a Russian chemist, M.S. Tswett, for separating and identifying the coloured pigments of plant leaves.

From the 1930s, a method called paper chromatography was used to analyse many more chemical substances, including dyestuffs, vitamins, fats, and blood pigments. A drop of liquid containing the substance to be analysed is placed at the bottom of a sheet of porous paper. The different parts or components of the mixture then creep up the paper at different rates, so that they separate out from one another. An even more recent invention is the gas chromatograph, which can analyse an even wider range of substances.

Spectrochemical analysis is another powerful method of identifying what a substance contains. It has a history going all the way back to Isaac Newton's discovery of the spectrum (see pages 34-35). Newton discovered that white light can be split up by a glass prism into a number of coloured bands. In

a similar way, the light from every chemical element, when this is made white-hot, will separate out into bands which identify that element.

In 1859 an instrument called the spectroscope was invented for this purpose by two Germans, Gustav Kirchhoff (1824-1887) and Robert Bunsen (1811-1899). Spectroscopes are not only used in the chemical laboratory. In astronomy, they analyse the light coming from stars and other parts of outer space, telling us what chemical elements and compounds exist there.

Lasers and Fibre Optics

A laser is an instrument that produces an intense beam of coherent light, that is, light of a single wavelength, where all the waves are "in step". (By contrast, light from an ordinary electric bulb is of many wavelengths, not in step.) The first working laser was built in 1960 by the inventor Theodore Maiman in the USA.

Lasers have an increasing number of applications. Their intense beams of light extend a very long way, so that they can be used for signalling. More frivolously, criss-crossing beams of laser light are sometimes used to make spectacular, brightly-

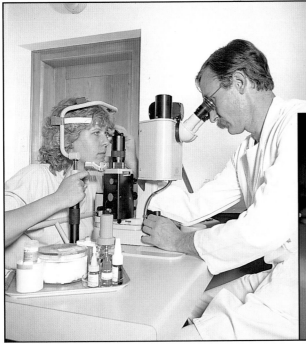

Left and below: Versatile lasers: their high-energy light beams are used for such widely different purposes as welding back a detached retina in the human eye and cutting out complex shapes from a steel sheet.

coloured displays against the night sky.

The concentrated energy of laser beams can also take the form of heat. This leads to such different applications as welding back the retina of the human eye after it has become detached, and use in weapons designed to explode enemy missiles during war in space or "Star Wars". A laser is also part of the compact disc recorder (see pages 62-63).

Fibre optics also uses light waves. It employs long glass fibres, each composed of two different layers of glass. Light shone in one end of these fibres passes mainly *along* the fibres, because the outer layer of glass refracts the light waves in to the inner layer of glass. Very little light escapes sideways.

Fibre-optics instruments contain ropes or bundles of these lighted-up fibres, which allow someone looking down one end to see what is happening at the other. A surgeon may use such an instrument to look inside a patient's body, and even carry out internal operations without the need to open the patient up.

Messages can also be sent along optic fibres, and fibre-optic cables now span both the Atlantic and Pacific Oceans.

Right: A fibre optics cable contains thousands of long glass fibres, along which light passes. One use of fibre optics is to send picture messages.

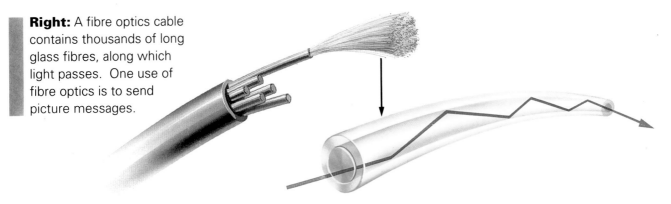

MAIMAN INVENTS LASER 1960 • FIBRE OPTICS USED FOR OPERATIONS 1970

COMPUTERS

People have always looked for quick ways to work out sums and have invented machines for doing so. One of the oldest calculating machines is the abacus, on which the answers to sums are found by moving beads on rods or wires (see pages 14-15).

Thousands of years later than the abacus, further calculating machines were invented, the latest of which are the computers we use today. To "compute" a sum means the same as to "calculate" it.

Analogue and Digital Computers

There are two basic types of computer or calculator. One type is called digital, because such machines work directly with numbers, or digits. Both the ancient abacus and the modern desk computer are examples of digital computers.

The other type is called analogue, because these machines do not work directly with numbers but with some other quantity that varies and can be measured: an analogous quantity. One example of an analogue computer is the slide rule, invented by an Englishman, John Gunter, in 1620. This employs lengths as the analogue to numbers. Other, more complicated analogue computers are used in industry for measuring and controlling processes such as the rate of flow of a product in a factory.

Early digital computers, or calculators, were mechanical machines using interlocking toothed wheels. One wheel would rotate ten times to turn another once, and by such means simple arithmetical sums could be quickly performed. The French philosopher Blaise Pascal (1623-1660) invented a calculating machine of this sort in 1642, which was used for gambling calculations.

In the 1820s an Englishman, Charles Babbage (1792-1871), invented a mechanical calculating machine that would carry out even more complicated arithmetic. He went on to design still more advanced machines, but none of these "difference engines" was completed.

Modern Computers

The "father" of today's computers may be said to be Alan Turing (1912-1954), an English mathematician who invented a theoretical digital computer in 1937. In 1944, the first such computer was built in the USA by Howard Aiken. Named the Harvard Mark 1, it was more than 15 m in length and filled a large room.

Equally large but more powerful was a computer built in 1946 by J.G. Brainerd and others at the University of Pennsylvania, USA. This was the first truly electronic computer. It employed vacuum tubes or radio valves (see pages 62-63) and could multiply 300 numbers each second. It printed out its answers on punched cards, a system invented in 1890 by Herman Hollerith in the USA.

The next major advance in computing was the replacement of bulky radio valves by much smaller

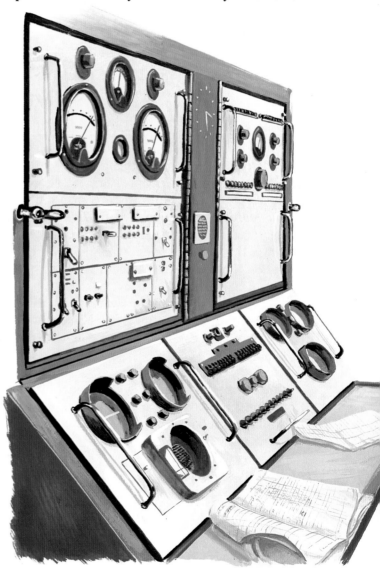

transistors. The transistor was invented in 1945 by William Shockley, John Bardeen and others in the USA. In another advance, electronic printers replaced the older punched card printers.

Computers were shrunk even further by printed circuits, developed after 1945. All electrical wiring was replaced by tiny electrical circuits deposited or printed on non-conducting boards. In 1958 computers were shrunk yet again, when the first integrated circuit was invented. Integrated circuits are single chips of a hard non-metal, silicon, on which highly complicated electrical circuits are micro-miniaturized. A microchip of today, less than 4 mm square, may contain millions of transistors.

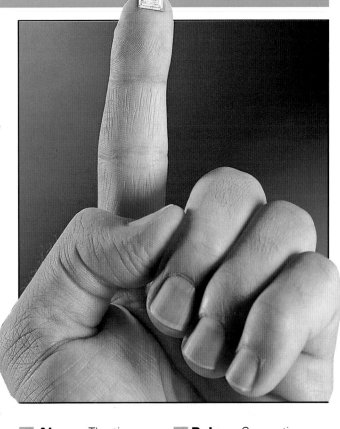

Below: An early electronic computer of the 1950s filled a whole room. Today, even a pocket calculator is more powerful!

Above: The tiny but powerful silicon chip which is used in computers of today.

Below: Computing can be carried out almost anywhere, using a laptop or desktop machine.

PRINTED CIRCUITS INVENTED 1945 • INTEGRATED CIRCUITS 1958

SPACE EXPLORATION

Space Pioneers

Konstantin Tsiolkovksy (1857-1935) of Russia became the father-figure of modern space exploration, when in 1903 he worked out the theory of spaceflight. He also invented a wind tunnel for testing the effects of rapid flight through air, such as those encountered by a rocket vehicle returning from space.

In 1935 Robert Goddard (1882-1945) in the USA, invented a liquid-fuelled rocket that flew faster than the speed of sound, though not into space. Hermann Oberth in Germany had invented a similar rocket four years earlier, and went on to invent a solid-fuelled rocket in the 1950s.

During World War II, Oberth worked with Wernher von Braun (1912-1977) on the invention of the *V2* war rocket. This stood 14 m high and had a rocket motor fuelled with alcohol and liquid oxygen. On its way to devastate parts of southern England, it rose to a maximum height of 100 km, not as far as outer space.

Above: The Space Shuttle being carried piggyback on a Boeing 747.

The Space Race

In 1949 a team in the USA, including Wernher von Braun, launched a rocket to a height of 400 km - well into space. The first man-made object to orbit the Earth, in 1957, was the artificial satellite *Sputnik 1* from the USSR. *Explorer 1*, the USA's first satellite, followed in 1958.

The first man in space, in 1961, was the Soviet cosmonaut Yuri Gagarin (1934-1968) orbiting in his spacecraft *Vostok 1*. John Glenn from the USA followed him into orbit the next year. The first man to "walk" in space was another Soviet, Alexei Leonov, who in 1965 spent 20 minutes attached to a cable outside his spacecraft, *Voshkod 2*.

Up to this time the Soviets were winning the space race, but the Americans surged ahead with the

V2 ROCKET 1944 • *SPUTNIK 1* ORBITS EARTH 1957 • FIRST MAN ON MOON 1969

Left: On its way to the edge of the Solar System, a *Voyager* space probe flies by the giant planet Saturn and its moons.

Below: Such space vehicles as HOTOL (Horizontal Take-Off and Landing) will succeed the Space Shuttle by the early twenty-first century.

Apollo Project. This used giant *Saturn* liquid-fuelled rockets, 111 m high, with a thrust of 3.5 million kg. In 1969 Neil Armstrong stepped down from the moon lander of *Apollo 11* to become the first human being to set foot on another world.

Space Explorers and Stations

In 1967 the Soviets launched their first space station or laboratory, *Soyuz 1*. The Americans followed this with *Skylab* in 1973. These and later space stations, manned by relays of scientists, orbit the Earth for extended periods.

Many other, unmanned, satellites have been launched to carry out automatic tasks of scientific observation. These include observation of the weather and exploration of the Earth's mineral, agricultural, and marine resources. Other satellites relay radio and TV programmes and messages around the entire globe.

Many unmanned space probes have travelled out to explore the Sun's family of planets. Those sent from the USA include *Mariner* space probes to Mars, Venus and Mercury, 1962-1975; *Viking* landers on Mars, 1975; and *Voyager 2* which in 1989, after a space journey of 12 years, flew past Neptune.

In 1981 the USA launched *Columbia*, its first manned Space Shuttle vehicle, which returned to Earth after orbiting it at a height of 480 km. With its huge drop tank and two solid rockets, the Space Shuttle blasts off with a thrust of 2 million kg. Later Shuttles have been used to launch, in 1989, *Hubble*, the first large space telescope, and in 1990, *Ulysses*, a space probe that will explore the Sun.

MARINER AND *VOYAGER* PROBES 1962-1989 • SPACE SHUTTLE LAUNCHED 1981

INDEX

ACKNOWLEDGEMENTS

The publishers would like to thank the following organizations and individuals for their kind permission to reproduce the pictures in this book:

AEA Technology, Harwell 69; Aberdeen City Arts-Libraries 56; British Aerospace (Space Systems) Ltd, Stevenage 77 bottom; J. Allan Cash Photolibrary, London 7, 39, 59 top, 61 left, 67, 73 right; Bruce Coleman, Uxbridge/Peter Davey 6; C.M. Dixon, Kingston, Canterbury 15 right, 16 bottom, 23; Dixons Ltd 63; Mary Evans Picture Library, London 44 top; Werner Forman Archive, London 19; Michael Holford, Loughton 11, 12 top, 12 bottom, 15 left, 16 top, 20, 25 left, 25 right, 26, 29, 32 left, 32 right, 35 left, 35 right, 43 top; Kodak Museum 53; Mansell Collection, London 66; Panos Pictures/Tom Learmonth 8,/Bruce Paton 44 bottom; QA Photos, Hythe 43 bottom; Ann Ronan Picture Library, Bishops Hull 38, 40, 46-47, 51, 62; Simon-Carves Ltd 70, 71; TRH Pictures, London 61 right; Ron Taylor 37; John Watney Photo Library, Bowness on Solway 57; Zefa Picture Library, London 45, 59 bottom, 65, 73 left, 75 top, 75 bottom, 76, 77 top and front cover.

Illustrations by:

Ray Hutchins - pages 21, 32-33, 40-41, 60 top
Joe Lawrence - pages 36, 57, 58, 69
The Maltings Partnership - pages 14, 22, 25, 27, 38, 42, 49, 51, 52-53, 53, 54-55, 60 middle, 60-61, 62-63, 65, 67, 68, 72, 73
Russell and Russell Associates - pages 30-31
Mark Stacey - title page, pages 6-7, 8, 10, 13, 17, 18-19, 24, 28-29, 30 left, 34-35, 35, 37, 48, 50, 64, 74-75

All illustrations in top right-hand corners by Joe Lawrence, except: page 39 (Ray Hutchins), page 47 (The Maltings Partnership), page 61 (Mark Stacey), page 65 (Bill Le Fever).

The illustrations in the top right-hand corners of the right-hand pages in this book show the following:

Page 7, fire; page 9, early sickle; page 11, Stonehenge; page 13, Ancient Egyptian jewellery; page 15, early domestic scales; page 17, the Hebrew letters aleph and beth; page 19, pair of bellows; page 21, early wheel; page 23, the Pharos lighthouse; page 25, Vitruvian water wheel; page 27, Chinese clock made by Su Sung, 1090; page 29, Chartres Cathedral, France; page 31, photocopier; page 33, John Harrison's chronometer; page 35, early spectacles; page 37, electron microscope; page 39, Model T Ford; page 41, Cugnot's steam carriage; page 43, The Golden Gate Bridge, San Francisco, USA; page 45, spinning wheel; page 47, Jethro Tull's seed drill; page 49, car battery; page 51, early telephone; page 53, zoetrope; page 55, Elias Howe's sewing machine, 1846; page 57, hypodermic syringe; page 59, penny farthing; page 61, Leonardo da Vinci holding a model of his design for a rotating-wing aircraft; page 63, portable radio and cassette player; page 65, 1930s radio; page 67, donkey pump for extracting oil; page 69, uranium atoms splitting in a nuclear reaction known as the cascade effect; page 71, sticks of dynamite; page 73, seismograph; page 75, slide rule; page 77, moon buggy

Illustrations in the pull-out time chart show, from left to right:

Pyramids at Giza, Egypt; early coins; armillary sphere; early cannon, 1346; Leonardo da Vinci holding a model of his design for a rotating-wing aircraft; spinning jenny; the Montgolfier brothers' hot-air balloon; first motorcycle, designed by Ernest and Pierre Michaux; contact lens; vacuum cleaner, 1910; Biro pen; radio telescope; mushroom cloud following an atomic bomb explosion; pocket calculator; robots welding parts of a car; video recorder